I0486677

ACE GENETICS!

(THE EASY GUIDE TO ACE GENETICS)

BY: DR. HOLDEN HEMSWORTH

Copyright © 2015 by Holden Hemsworth

All rights reserved. No part of this publication may be reproduced, distributed, or transmitted in any form or by any means, including photocopying, recording, or other electronic or mechanical methods, without the prior written permission of the publisher, except in the case of brief quotations embodied in critical reviews and certain other noncommercial uses permitted by copyright law.

DISCLAIMER

Genetics, like any field of science, is continuously changing and new information continues to be discovered. The author and publisher have reviewed all information in this book with resources believed to be reliable and accurate and have made every effort to provide information that is up to date and correct at the time of publication. Despite our best efforts we cannot guarantee that the information contained herein is complete or fully accurate due to the possibility of the discovery of contradictory information in the future and any human error on part of the author, publisher, and any other party involved in the production of this work. The author, publisher, and all other parties involved in this work disclaim all responsibility from any errors contained within this work and from any results that arise from the use of this information. Readers are encouraged to check all information in this book with institutional guidelines, other sources, and up to date information.

The information contained in this book is provided for general information purposes only and does not constitute medical, legal or other professional advice on any subject matter. The author or publisher of this book does not accept any responsibility for any loss which may arise from reliance on information contained within this book or on any associated websites or blogs.

WHY I CREATED THIS STUDY GUIDE

In this book, I try to breakdown the content covered in most undergraduate Genetics courses in college for easy understanding and to point out the most important subject matter that students are likely to encounter. This book is meant to be a supplemental resource to lecture notes and textbooks to boost your learning and go hand in hand with your studying!

I am committed to providing my readers with books that contain concise and accurate information and I am committed to providing them tremendous value for their time and money.

Best regards,

Dr. Holden Hemsworth

TABLE OF CONTENTS

CHAPTER 1: INTRODUCTION TO GENETICS

What is Genetics?

Genetics is the scientific study of heredity and heredity variation. Heredity is the passing of biologicals traits from one generation to the next that results from the transmission of genes from parents to offspring.

- DNA (deoxyribonucleic acid), hereditary material in almost all organisms

    o Notable exception: hereditary material in retroviruses is RNA

- Chromosomes - long threadlike structure composed of chromatin

    o Carry genetic information in the form of genes

    o Contain hundreds or even thousands of genes at a locus

        ▪ Locus - specific location on a chromosome containing a gene

- Genes – units of hereditary information made of DNA

    o Most genes contain information to synthesize specific proteins

Fields of Genetics

Genetics is traditionally divided into three areas: transmission genetics, molecular genetics, and population genetics.

Transmission Genetics

- Examines how traits are transferred from parent to offspring

- Mendel provided the conceptual framework for this field

    o He showed that genetic determinants pass from parent to offspring as discrete units

Molecular Genetics

- Examines genes

    o Their features, organization, and function

    o As well as mutant genes with abnormal function

- Study model organisms

    o Model organisms are species that have been widely studied

    o Usually, they have particular experimental advantages

        ▪ Easy to breed, breed quickly, easy to maintain, etc.

Population Genetics

- Examines the genetic composition of populations (i.e., the distribution and changes of allele frequency in a population)

- Allele frequency is subject to four main evolutionary processes

    o Natural selection

    o Gene flow (migration)

        ▪ Migration moves alleles between populations

    o Mutation

    o Genetic drift

        ▪ Changes in allele frequencies due to random sampling from the population

Brief History of Genetics

- 1866 – Mendel published his ideas on inheritance, equal segregation, and independent assortment

    o Significance of his work was not recognized for over 30 years

- 1869 – DNA (initially called "nuclein") is identified by Friedrich Miescher

    o Significance of DNA was not realized for over 70 years

- 1900 – Mendel's work is rediscovered

- 1902 – Disease is first related with genetic causes, chromosome theory of inheritance proposed by Sutton

    o Theory identified chromosomes as carriers of genetic material

- 1944 – DNA is identified as the hereditary material

    o Skepticism remained for years to follow

- 1953 – Watson and Crick figured out that DNA is in the shape of a double helix

    o Determination of the structure of DNA led to rapid developments in molecular genetics

- 1970s – Gene cloning and DNA sequencing methods developed

- 1980s – DNA marker analysis methods, DNA fingerprinting methods, and PCR is developed

- 1990s – Whole genome sequencing methods developed leading to Genomics, Proteomics, and Bioinformatics

    o Genomics – study of all genes in a genome

    o Proteomics – study of all proteins coded by a genome

    o Bioinformatics – science of collecting and analyzing complex biological data

- 2003 – Human Genome Project is completed after being initiated in 1990

Charles Darwin and Natural Selection

Natural selection is the differential survival and reproduction of individuals that differ in phenotype. Adaptations are the result of the natural weeding out process of unfit individuals that is a part of natural selection (i.e. specific environmental pressures favor the reproductive success of some individuals over others).

Charles Darwin:

- Charles Darwin stated in his book, *On the Origin of Species,* that a mechanism of evolutionary change was natural selection

- Darwin came up with the idea of natural selection based on the following observations:

  - Individuals of a population vary in inheritable traits

  - Populations can potentially produce more offspring than the environment can support

  - Individuals best adapted to their environment reproduce more often

    - Increases the proportion of traits shared by these individuals in the next generation

- Natural selection increases the frequency of inherited variants that arise by chance

- Darwin proposed that new species could arise from ancestral ones

  - Due to cumulative changes in a population over long periods of time

Biomolecules

Biomolecules consist of carbon and hydrogen and are present in all living organisms. The four major types of biomolecules are:

Carbohydrates

- Also called saccharides

- Contain multiple -OH groups

- Readily convertible between open and close (cyclic) form

Carbohydrate Representation Examples

Nucleotides

- 5-carbon sugar

- Nitrogen containing base ring

- PO_3 group

Nucleotide Representation Example

Lipids

- Long chain of hydrocarbons

- Mostly nonpolar

    o Makes them water-insoluble, generally

- Can be amphipathic

    o Amphipathic - contain both polar and nonpolar components

- Steroids are lipids that have fused rings in their structure

Example of Cholesterol (a Steroid):

Lipid Example of Triglyceride:

Amino Acids

- All have at least 2 ionizable groups (an amino and a carboxyl group)

- Their identities are determined by the side chain (R group)

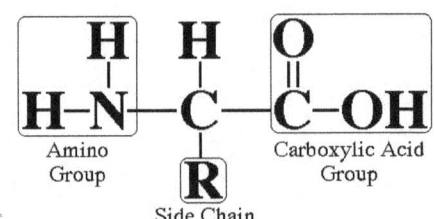

General Structure of Amino Acids

DNA

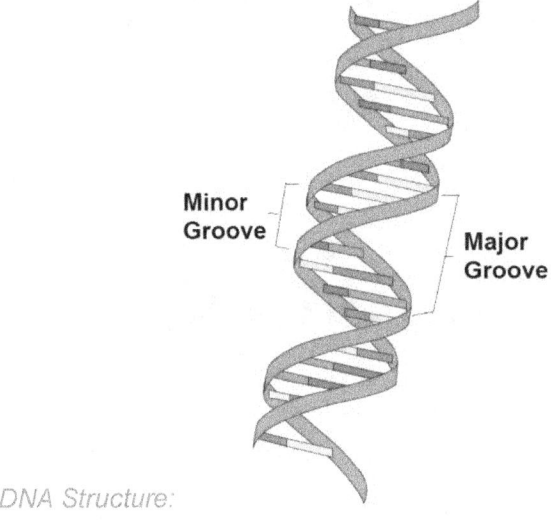

DNA Structure:

Structure of DNA

- Polymer of nucleotides

    o Adenine (A), thymine (T), cytosine (C), and guanine (G)

- Double helix

- DNA "backbone" is in a repeated pattern: deoxyribose, phosphate, deoxyribose, phosphate…

- Strands held together by hydrogen bonds between nitrogenous base pairs

    o A pairs with T

        ▪ Two hydrogen bonds between them

    o C pairs with G

        ▪ Three hydrogen bonds between them

- Two strands run antiparallel to one another
  - 5' to 3' pairs with 3' to 5'

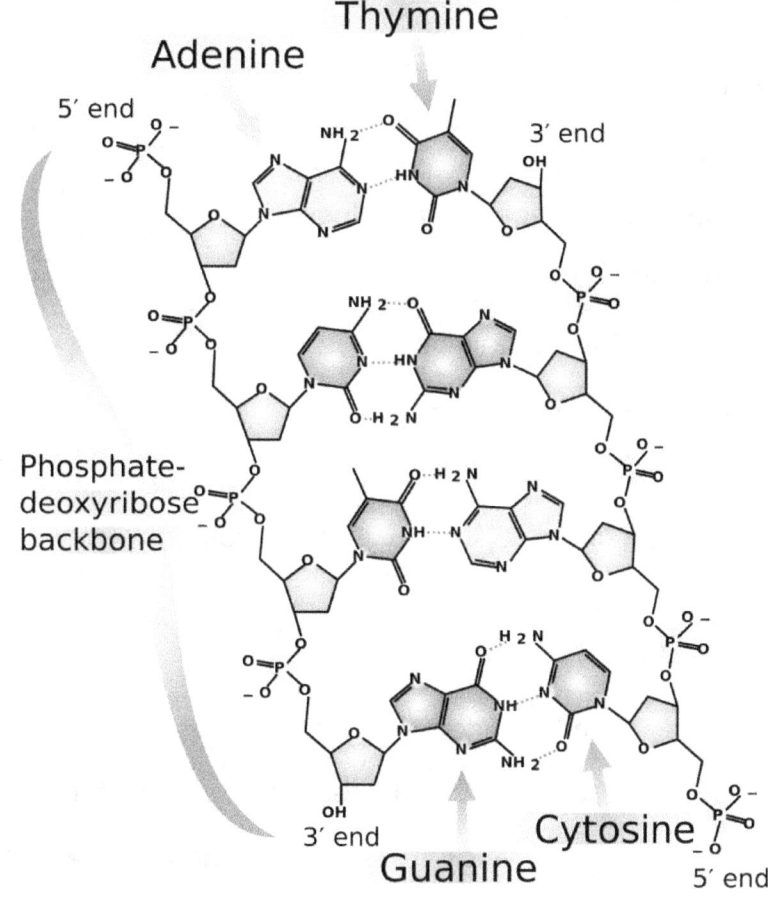

DNA Structure:

- Base pairs are perpendicular to the backbone
  - Base pairs stack closer together
  - Movement of bases together tilts the backbone by 30°
    - Creates an uneven twist
- Major groove vs. Minor groove
  - Grooves are spaces between the backbones
  - Major groove occurs where the backbones are far apart
  - Minor groove occurs where they are close together

Proteins

DNA holds the information for protein synthesis. Proteins are polymers made of amino acids. Characteristics of a cell are largely influenced by the proteins it produces. Proteins serve diverse biological functions.

Types of Proteins

- Binding proteins
  - Bind and release small molecules
- Structural proteins
  - Help a cell maintain its shape
  - Some help with movement
- Motor proteins
  - Convert chemical energy into physical movement
- Enzymes – biological catalysts
  - Catabolic enzymes
    - Help breakdown large molecules into smaller ones
    - Breakdown usually results in energy being released
      - Energy can then be used for other cellular activities
  - Anabolic enzymes
    - Help synthesize large molecules from smaller ones

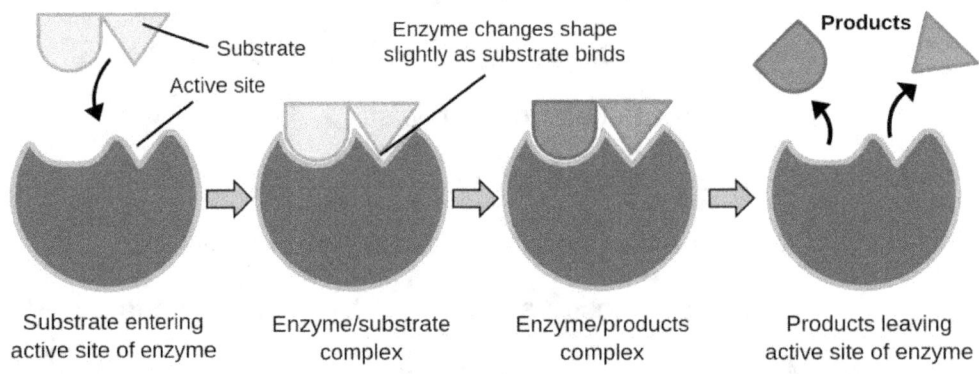

| Substrate entering active site of enzyme | Enzyme/substrate complex | Enzyme/products complex | Products leaving active site of enzyme |

Enzyme Catalysis:

The Central Dogma

The central dogma describes the flow of genetic information in cells.

- DNA → RNA → Protein
- DNA is replicated with each division cycle
  - Replication is semi-conservative
    - One parent strand is conserved
- DNA is transcribed into mRNA
  - Transcribe one form of nucleic acid into another form
- mRNA is translated into protein
  - Language of nucleic acid translated into language of amino acids
- Notable exceptions to the dogma are viruses

Traits

Traits are any characteristic that an organism displays. Molecular expression of genes within cells is responsible for the visible surface traits of an organism.

- Types of traits
  - Morphological
    - Affect appearance (e.g. color)
  - Physiological
    - Affect function (e.g. ability to metabolize lactose)
- Inherited differences in traits are attributed to genetic variation
  - Genetic variation – differences in inherited traits among individuals or within populations
- Genetic variation results from
  - Mutations
    - Small differences in the gene sequence
  - Change in chromosome structure
    - Segments of a chromosome may be lost or duplicated

- Change in chromosome number
  - Single chromosome may be lost or gained
  - Whole set of chromosomes may be inherited
- Traits are also influenced by the environment

Asexual vs. Sexual Reproduction

Asexual Reproduction

- Single individual is the only parent
- Parent passes all of its genes to its offspring
  - Offspring are genetically identical to the parent (clone)
    - Genetic differences can occur between offspring and parent from mutations

Sexual Reproduction

- Two parents produce an offspring
- Each parent passes half of their genes to the offspring
- Greater genetic variation is achieved
  - Offspring vary genetically from their siblings and parents
- Sexually-reproducing species are diploid
  - Diploid – contain two complete sets of chromosomes
    - One from each parent
- The two copies are called homologues
  - Homologues contain the same gene
    - May not contain the same alleles
    - Allele - one of two or more alternative forms of a gene located on the same place on a chromosome
- Human somatic (body) cells have 46 chromosome
  - 23 homologous pairs

- Gametes – haploid reproductive cells
    - Haploid – contain one set of chromosomes
    - Sperm and ovum are gametes
    - Human haploid number is 23
        - Diploid number is restored when two haploid gametes unite during fertilization
        - Fertilization - union of two gametes to form a zygote
        - Zygote - diploid cell resulting from the union of two haploid gametes

CHAPTER 2: CLASSICAL GENETICS – MENDELIAN INHERITANCE

Important Genetic Terminology

- Homozygous - having two identical alleles for a given trait (e.g., AA or aa)
    - Homozygotes are true-breeding
- Heterozygous - having two different alleles for a trait (e.g., Aa).
    - Heterozygotes are not true-breeding
- Phenotype - an organism's expressed traits
- Genotype - an organism's genetic makeup

Mendel's Experiments

Mendel is considered the founder of modern genetics. In 1857, Mendel was living in an Augustinian monastery where he bred garden peas. He chose them as his experimental organisms. From his experiments Mendel established many rules of heredity.

Gregor Mendel:

- Mendel chose characters in pea plants that were easy to distinguish
    - Character - detectable inheritable feature of an organism
    - Trait - variant of an inheritable character
- Chose seven characters, each of which had two traits
    - Flower color (purple or white)
    - Flower position (axial or terminal)

- Seed color (yellow or green)

- Seed shape (round or wrinkled)

- Pod shape (inflated or constricted)

- Pod color (green or yellow)

- Stem length (tall or dwarf)

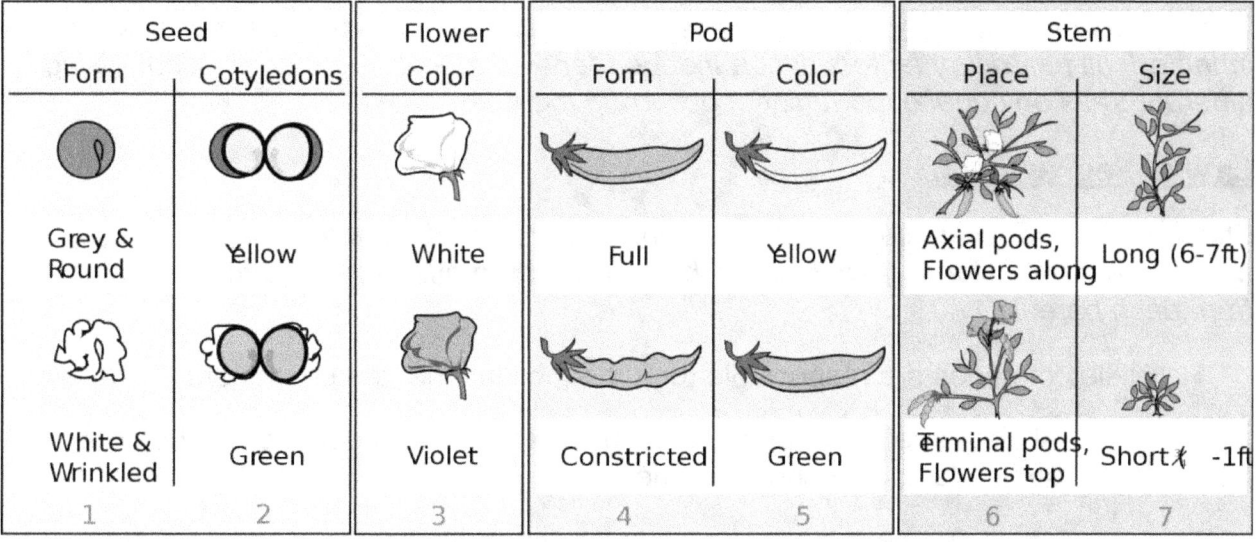

Seven Characteristics Studied by Mendel

- Mendel started his experiments with true-breeding plants and hybridized (cross-pollinated) them in experimental crosses

  - Hybridization – mating or crossing two individuals with different characteristics

    - Example: mating a purple-flowered plant with a white-flowered plant

  - Hybrids – offspring resulting from hybridization

  - True breeding – when the parents are self-fertilized they always produce offspring with the same traits as them

    - True-breeding parental plants of such a cross are called the P generation

    - Hybrid offspring of the P generation are called the F_1 generation

    - Allowing F_1 generation plants to self-pollinate produces the F_2 generation

- Mendel observed selected traits for at least three generations
  - Arrived at two principles of heredity
    - The law of segregation
    - The law of independent assortment

Law of Independent Assortment

The law of independent assortment states that when two or more characteristics are inherited, all hereditary factors assort independently during gamete production, giving different traits an equal opportunity of occurring together.

Law of Segregation

The law of segregation states that during the production of gametes the two copies of each hereditary factor segregate in such a way that the offspring acquire one factor from each parent.

- Alleles of genes are responsible for variations in inherited characters
  - Allele – one of two or more alternative forms of a gene located on the same place on a chromosome
  - Example: the gene for flower color in pea plants exists as two alleles
    - One for purple color and one for white color
- For each character, an organism inherits two alleles (one from each parent)
- If two alleles differ:
  - One is fully expressed (dominant allele)
    - Dominant alleles are designated by a capital letter
  - Other is completely masked (recessive allele)
    - Recessive alleles are designated by a lowercase letter
- The two alleles for each character segregate during gamete production
- Law of segregation predicts a 3:1 phenotypic ratio observed in the F_2 generation of a monohybrid cross
  - Combinations resulting from a genetic cross can be predicted using a Punnett square

o Example cross between two heterozygotes:

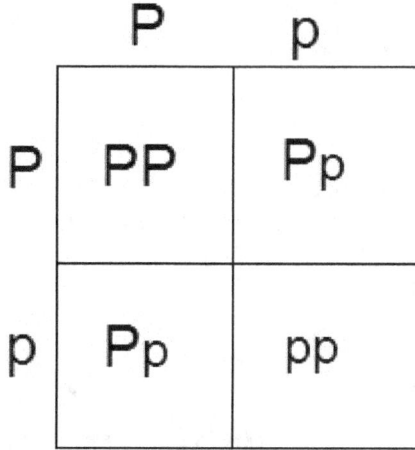

Punnett Square:

o P = allele for purple flower (dominant allele)

o p = allele for white flower (recessive allele)

o The F_2 progeny include:

- ¼ of the plants with two alleles for purple flowers (PP)

- ½ of the plants with one allele for purple flowers and one allele for white flowers (Pp)

    • Since the purple-flower allele is dominant, Pp plants will have purple flowers

- ¼ of the plants with two alleles for white flower color (pp)

Testcross

A testcross is the breeding of an organism of unknown genotype with a homozygous recessive. Since some alleles are dominant, the genotype of an organism may not always be obvious.

- Using the example from above; a pea plant with purple flowers could either be homozygous dominant (PP) or heterozygous (Pp)

- To determine whether an organism is homozygous dominant or heterozygous a testcross is utilized

- For example: if a cross between a purple-flowered plant of unknown genotype (P_) produced only purple-flowered plants, the parent was probably homozygous dominant

    o PP x pp cross produces all purple-flowered progeny that are heterozygous (Pp)

- If the progeny of the testcross contains both purple and white phenotypes, then the purple-flowered parent was heterozygous
    - Pp x pp cross produces Pp and pp progeny in a 1:1 ratio

Forked-line Method

The forked-line method is a way of calculating the predicted ratios of offspring by multiplying the probabilities of independent events. Especially useful method for when crosses start getting complicated. Example of using the forked-line method:

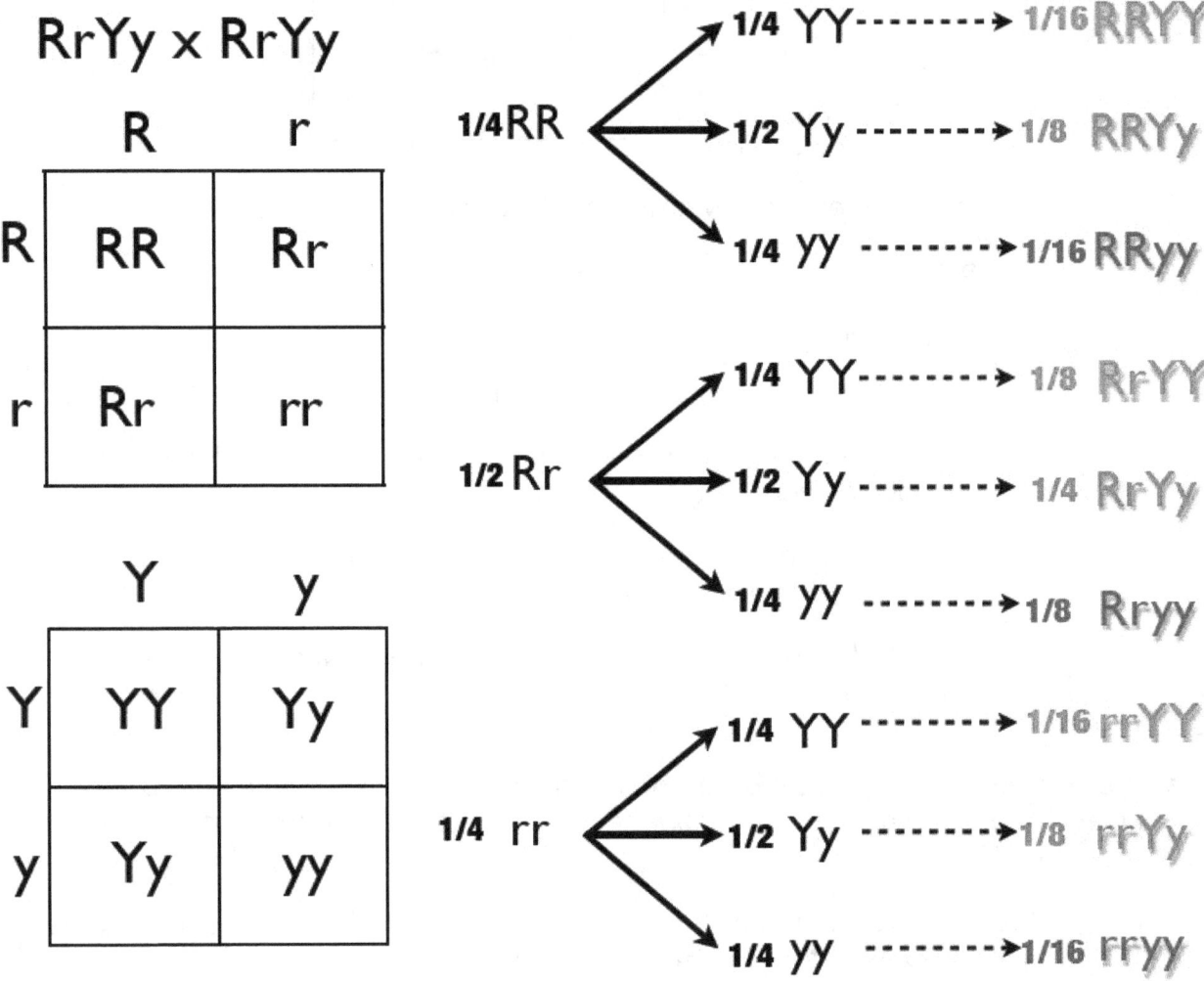

Inheritance in Humans

When studying human traits, there are no controlled parental crosses, so we have to rely on information from family trees or pedigrees. Pedigree analysis can be used to determine the pattern of inheritance of particular traits in humans.

- Pedigree - a family tree that diagrams the relationships among parents and children across generations

  - Also shows the inheritance pattern of a particular phenotypic character

- Convention for a pedigree:

  - Squares represent males

  - Circles represent females

  - Diamonds represent an individual with unknown sex or unspecified sex

  - Diagonal lines through squares or circles represent deceased individuals

  - Horizontal line connecting a male and female indicates mating between them

    - Offspring are listed below their parents in birth order (left to right)

  - Shaded symbols indicate individuals with the trait of interest

  - Unshaded symbols indicate unaffected individuals (i.e. individuals without the trait of interest)

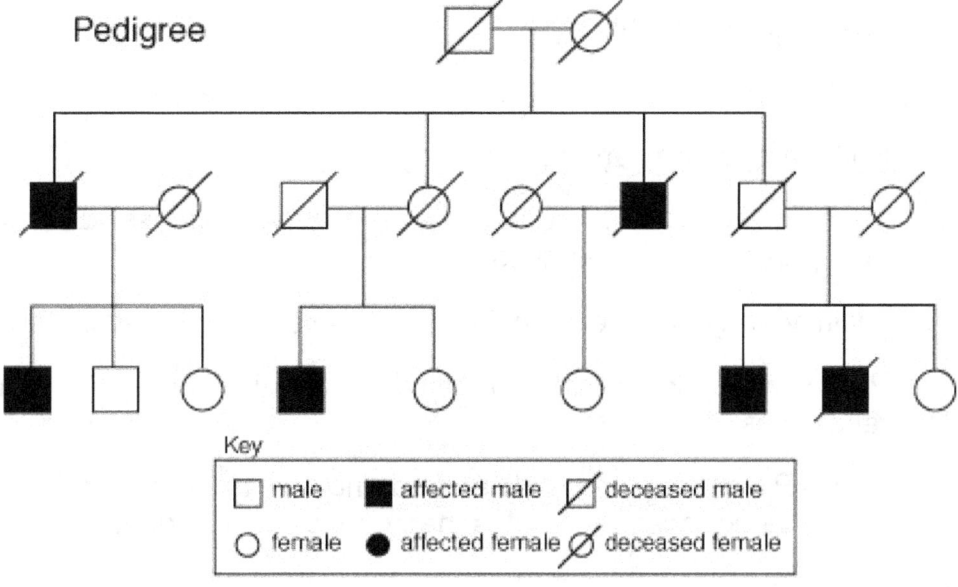

- Dominant Trait Pedigree (Widow's Peak Example)
  - If a widow's peak results from a dominant allele, W, then:
    - All individuals that do not have a widow's peak must be homozygous recessive (ww)
    - Individuals that have a widow's peak hairline must be either homozygous dominant (WW) or heterozygous (Ww)
- Recessive Trait Pedigree (Attached Ear Lobe Example)
  - If attached earlobes result from a recessive allele (f), then:
    - All individuals with attached earlobes must be homozygous recessive (ff)
    - Parents of an individual with attached earlobes must be either heterozygous (Ff) or have attached earlobes themselves (ff)
      - Neither parent can be homozygous dominant as all of their offspring would have a dominant allele

Recessively Inherited Disorders

Recessive alleles that cause human disorders are usually defective versions of normal alleles; they code for a malfunctioning protein or may not code for a protein at all.

- Heterozygotes might be phenotypically normal
  - If a single copy of the dominant allele produces sufficient quantities of the protein coded by the gene
- Most people with recessive disorders are born to normal parents
  - Both parents are carriers (heterozygous)
  - There is ¼ chance of producing a homozygous recessive zygote when both parents are heterozygous
- Consanguinity - a genetic relationship that results from shared ancestry
  - Probability is higher that consanguinous mating will result in harmful recessives
    - Parents with recently shared ancestry are more likely to inherit the same recessive alleles than unrelated individuals

Dominantly Inherited Disorders

- Homozygous dominant individuals are typically spontaneously aborted when they are a fetus in dominantly inherited disorders

    o Homozygous recessive individuals are considered the normal phenotype in these cases

- Lethal dominant alleles are much rarer than lethal recessives

    o Since they are always expressed when inherited

    o Usually result from new genetic mutations that occur in gametes and later kill the developing embryo

- Late-acting lethal dominants can escape elimination if the disorder does not take effect until the affected individual has reached a reproductive age

- Example: Huntington's disease is a degenerative disease of the nervous system

    o Phenotypic effects do not appear until 35 to 40 years of age

        ▪ Affected individual may already have offspring before signs of the disease show themselves

    o Gene for Huntington's has recently been identified to be located on chromosome #4

    o Children of an afflicted parent have a 50% chance of inheriting the lethal dominant allele

Multifactorial Disorders

- Multifactorial disorders are caused by both genetic and environmental factors

- Hereditary component is often polygenic

CHAPTER 3: CELL CYCLE AND MITOSIS

Human Life Cycle

- Somatic cell - any cell other than reproductive cells (sperm or egg cell)

    o Human somatic cells contain 46 chromosomes

    o Chromosomes vary in:

        ▪ Size, position of the centromere, and staining or banding pattern

- Homologous chromosomes (homologues) - pair of chromosomes that have the same size, centromere position, and banding pattern

    o Homologous autosomes carry the same genetic loci

    o Autosome - chromosome that is not a sex chromosome

- Sex chromosome - chromosomes that determine an individual's sex

    o Human sex chromosomes carry different loci

        ▪ Even though they pair during prophase of meiosis I

    o Human females have a homologous pair of X sex chromosomes (XX)

        ▪ Are called homogametic

    o Human males have one X and one Y chromosome (XY)

        ▪ Are called heterogametic

        ▪ Y chromosome determines maleness

        ▪ Males can only receive their Y chromosome from their father

- Humans have 22 pairs of autosomes and one pair of sex chromosomes

    o 23 pairs or 46 individual chromosomes in total

Cell Division

Cell division is a regulated process that results in the distribution of identical hereditary material to two daughter cells.

- In unicellular organisms, the division of one cell reproduces an entire organism

- In multicellular organisms, cell division allows:
  - Growth and development from the fertilized egg
  - Replacement of damaged or dead cells
- Eukaryotic chromosomes are supercoils of a DNA-protein complex called chromatin
  - Each chromosome consists of the following:
    - A single double-stranded molecule of DNA
    - Various proteins that serve to maintain the structure of the chromosome
      - Or involved in the expression of genes, DNA replication, DNA repair
- Whole genome is duplicated in preparation for eukaryotic cell division
  - Each chromosome consists of two sister chromatids following duplication
  - Two chromatids are initially attached to each other at a region called the centromere
- Cell division usually proceeds in two sequential steps
  - First nuclear division (mitosis) and then division of the cytoplasm (cytokinesis)
    - Not all cells undergo cytokinesis following mitosis
  - In mitosis sister chromatids are pulled apart and separated
    - One set at each end of the cell
  - In cytokinesis, the cytoplasm is divided and two separate daughter cells are formed
    - Each contains a single nucleus with one set of chromosomes

Animal Cell Cycle Diagram:

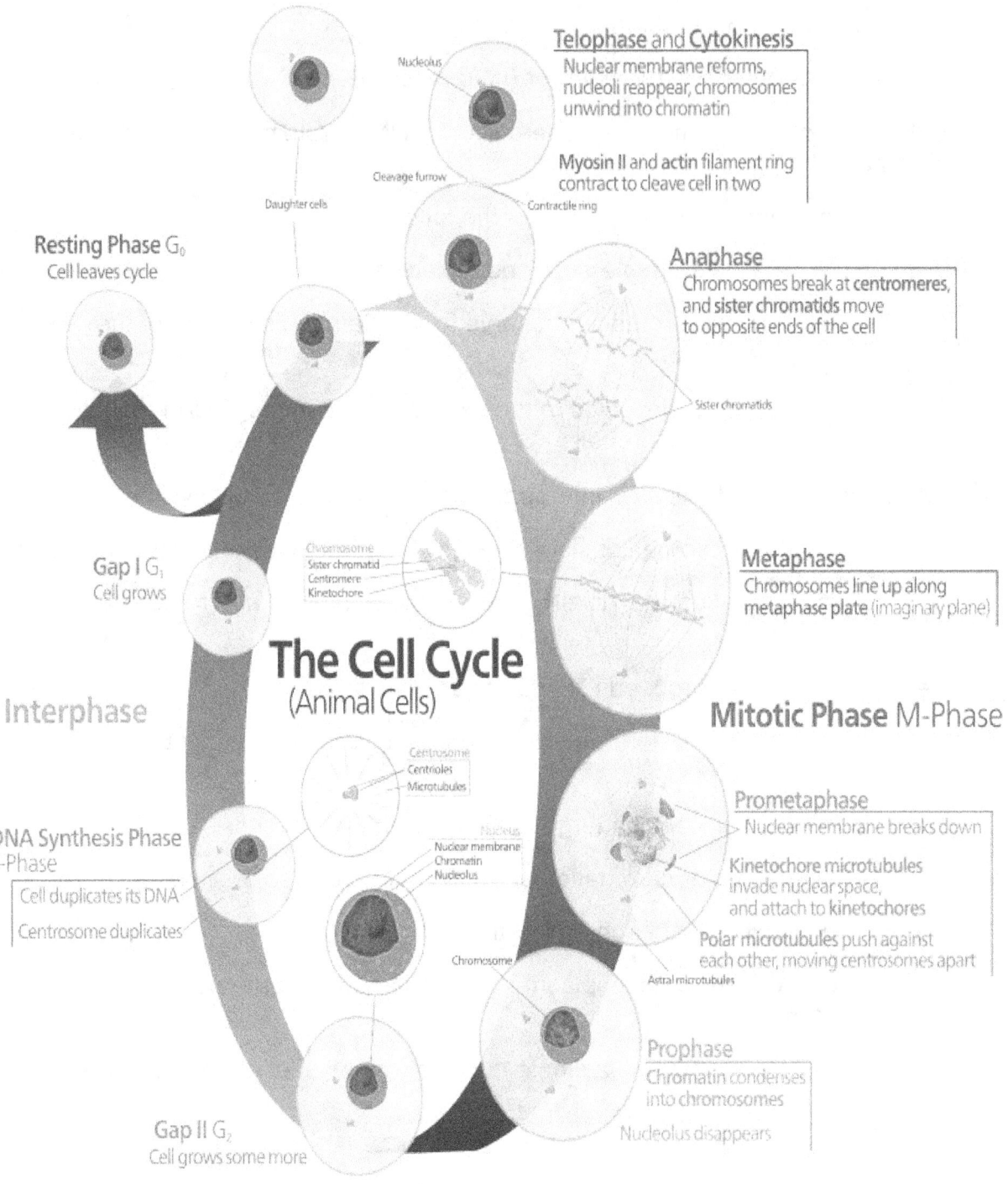

Mitotic Cell Cycle

The cell cycle occurs in an ordered sequence of events in which a cell duplicates its contents and then divides in two. Some cells go through repeated cell cycles, while other cells never or rarely divide once they are formed (e.g. muscle cells).

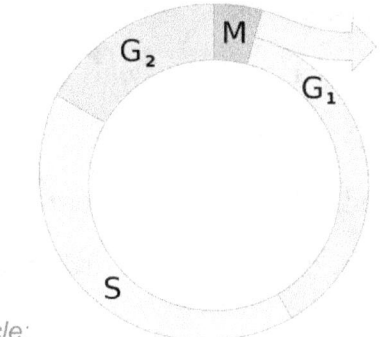

Cell Cycle:

- Cell cycle alternates between
    - Mitotic (M) phase - dividing phase
    - Interphase - non-dividing phase
- M phase is the shortest part of the cell cycle
    - The phase in which the cell divides
    - Includes both mitosis and cytokinesis
- Interphase is the period in which the cell grows and copies its chromosomes in preparation for cell division
    - ~90% of the cell cycle
- Interphase consists of three periods:
    - G_1 phase - First growth phase
        - Cell prepares to divide
        - Reaches a restriction point
        - Commits itself to cell division pathway
    - S phase - DNA is synthesized as chromosomes are duplicated
        - Chromatids – two copies of a replicated chromosome
        - Sister chromatids – chromatids joined by a centromere
    - G_2 phase - Second growth phase (growth and preparation for mitosis)

- Mitosis only occurs in eukaryotes
- Mitosis is a continuous process but it is typically divided into five stages
  - Prophase, prometaphase, metaphase, anaphase, and telophase
  - When cytokinesis occurs
    - We usually associate it with telophase of mitosis
- A cell may remain in G_0 phase indefinitely
  - Either the decision to divide has been postponed
  - Or the cell has decided to never divide again (i.e. terminally differentiated cells)

G_2 of Interphase

- G_2 cells are characterized by:
  - Well-defined nucleus bounded by a nuclear envelope
  - One or more nucleoli
  - Two centrosomes adjacent to the nucleus
  - Duplicated chromosomes cannot be distinguished individually due to loosely packed chromatin fibers
    - Chromosomes are duplicated earlier in the S phase
- In animal cells, further characterized by:
  - Pair of centrioles in each centrosome
  - Radial microtubule array around each pair of centrioles

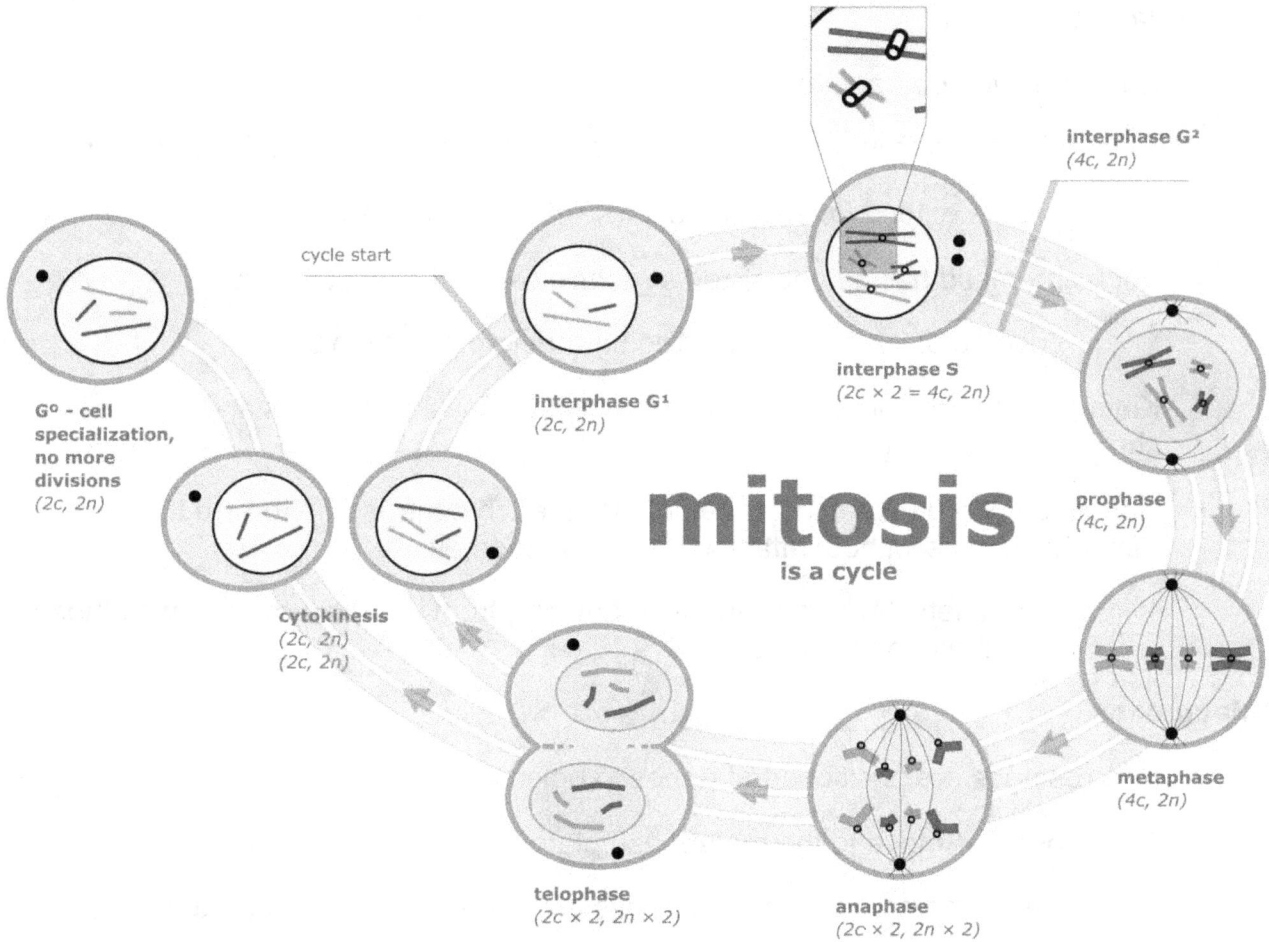

interphase G² *(4c, 2n)*

cycle start

interphase S *(2c × 2 = 4c, 2n)*

interphase G¹ *(2c, 2n)*

G⁰ - cell specialization, no more divisions *(2c, 2n)*

prophase *(4c, 2n)*

mitosis
is a cycle

cytokinesis *(2c, 2n)* *(2c, 2n)*

metaphase *(4c, 2n)*

telophase *(2c × 2, 2n × 2)*

anaphase *(2c × 2, 2n × 2)*

Prophase

- In the nucleus:

    - Nucleoli disappear

    - Chromatin fibers condense chromosomes

        - Composed of two identical sister chromatids joined at the centromere

- In the cytoplasm:

    - Mitotic spindle forms

        - Composed of microtubules between the two centrosomes

    - Centrosomes move apart due to the lengthening of the microtubule bundles between them

Prometaphase

- Nuclear envelope fragments

    o Allows microtubules to interact with the highly condensed chromosomes

- Spindle fibers extend from each pole toward the cell's center

- Each chromatid has a specialized structure called the kinetochore

    o Located at the centromere region during prometaphase

- Kinetochore microtubules become attached to the kinetochores and put the chromosomes in motion

- Non-kinetochore microtubules radiate from each centrosome toward the metaphase plate without attaching to chromosomes

    o Non-kinetochore microtubules radiating from one pole overlap with those from the opposite pole

Metaphase

- Centrosomes are positioned at opposite poles of the cell

- Chromosomes move to the metaphase plate (i.e. plane between spindle poles)

    o Centromeres of all chromosomes are aligned at the metaphase plate

- Kinetochores of sister chromatids face opposite poles

    o Identical chromatids are attached to kinetochore fibers radiating from opposite ends of the parent cell

    o Entire structure formed by non-kinetochore microtubules plus kinetochore microtubules is called the spindle

Anaphase

- Begins when paired centromeres of each chromosome start to separate

    o Sister chromatids split apart into separate chromosomes

        ▪ Move toward opposite poles of the cell

- Kinetochore microtubules shorten at the kinetochore end as chromosomes approach the poles

    o Simultaneously, poles of the cell move farther apart

- End of anaphase, the two poles have identical collections of chromosomes

Telophase and Cytokinesis

- Telophase

    o Non-kinetochore microtubules elongate the cell

    o Daughter nuclei begin to form at the two poles

    o Nuclear envelopes form around the chromosomes

        ▪ Form from fragments of the parent cell's nuclear envelope

        ▪ Also portions of the endomembrane system

    o Nucleoli reappear

    o Chromatin fiber of each chromosome uncoils and the chromosomes become less distinct

- Cytokinesis

    o Occurs in animal cells through a process called cleavage

    o Cleavage furrow forms (a shallow groove in the cell surface)

        ▪ Near old metaphase plate

    o A contractile ring of actin microfilaments forms

        ▪ Contracts until it pinches the parent cell in two

    o Remaining mitotic spindle breaks and the two cells are separated

Binary Fission

Mitosis, which only occurs in eukaryotes, may have evolved from binary fission in bacteria. Binary fission is a form of asexual reproduction in which bacteria replicate their chromosomes and distribute them between two daughter cells.

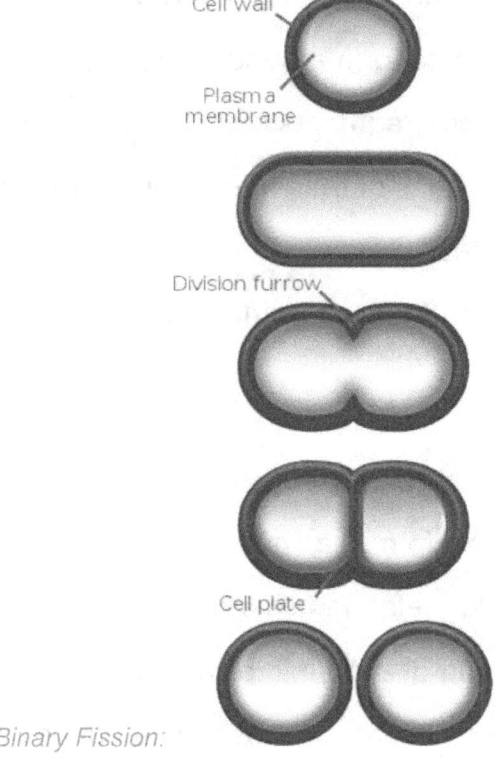

Binary Fission:

- Prokaryotes contain a single circular chromosome composed of a double-stranded DNA molecule

    o Reproduce by binary fission

- Chromosome is replicated and each copy remains attached to the plasma membrane at adjacent sites

    o Between these sites the membrane grows and separates the two copies of the chromosome

- Bacterium grows, plasma membrane pinches inward

- Cell wall forms across the bacterium between the two chromosomes

    o Divides the original cell into two daughter cells

CHAPTER 4: MEIOSIS AND SEXUAL REPRODUCTION

Sexual Reproduction

Sexual reproduction is the most common way eukaryotic organisms produce offspring.

- Parent make gametes which contain half the amount of genetic material

  - Gametes – haploid reproductive cells

  - Haploid – contain one set of chromosomes

  - Sperm and ovum are gametes

    - Sperm cells are relatively small and mobile

    - Egg cell or ovum are usually larger and non-mobile

      - In animal species, store a large amount of nutrients

- Gametes are 1n, while diploid cells are 2n

  - Diploid human cell contains 46 chromosomes

  - Haploid human cell contains 23 chromosomes

    - The diploid number is restored when two haploid gametes unite in the process of fertilization

    - Fertilization - union of two gametes to form a zygote

    - Zygote - diploid cell that results from the union of two haploid gametes

Meiosis

Meiosis and sexual reproduction significantly contribute to genetic variation among offspring. Meiosis includes steps that closely resemble steps observed in mitosis.

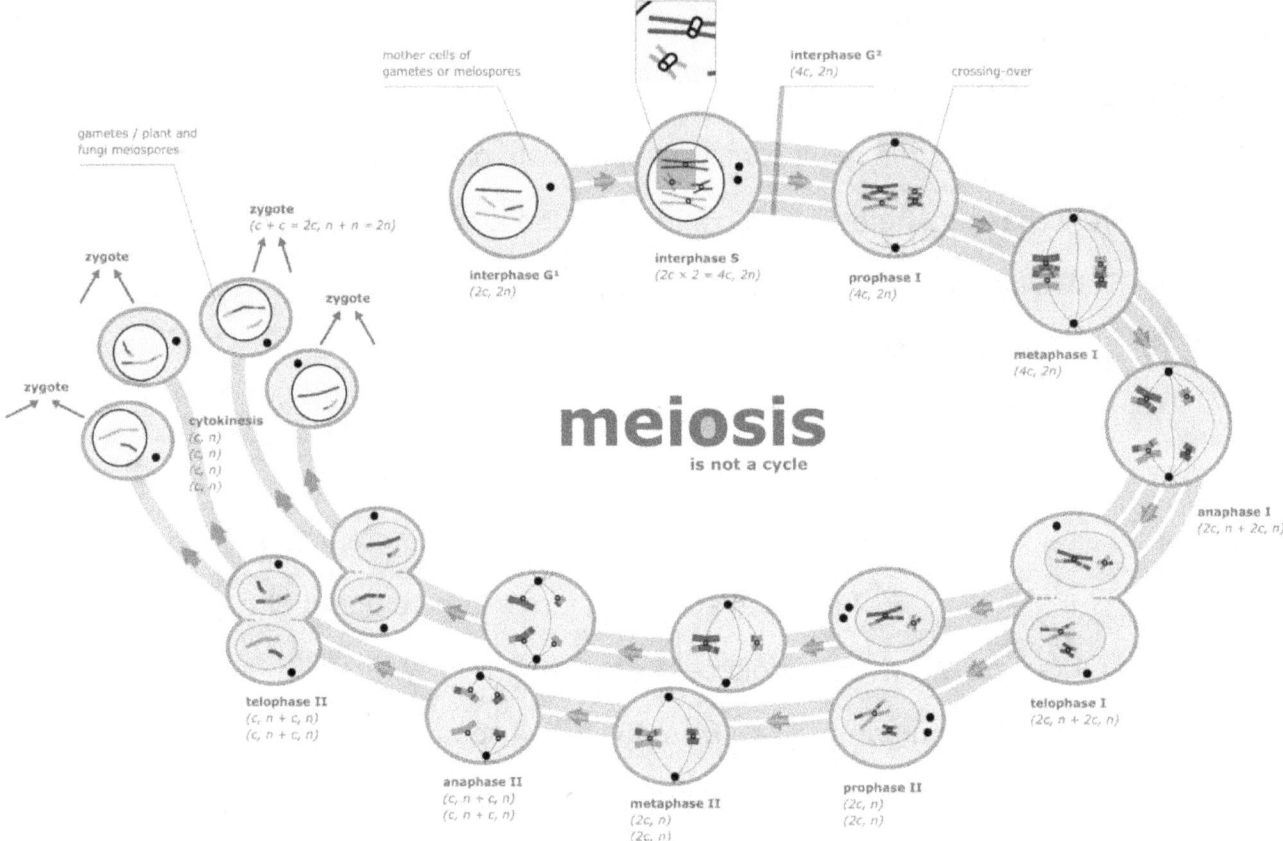

- Meiosis is preceded by replication of chromosomes

  - Meiosis differs from mitosis in that the single replication is followed by two consecutive cell divisions (meiosis I and meiosis II)

    - Two cell divisions produce four daughter cells

  - Daughter cells in **meiosis** have half the number of chromosomes as the original cell

    - Daughter cells of **mitosis** have the same number of chromosomes as the parent cell

Stages of Meiosis I

Keep in mind that there are two cell divisions in meiosis, meiosis I and meiosis II. Meiosis I divides the two chromosomes of each homologous pair and reduces the chromosome number by ½.

- Interphase I - precedes meiosis

  - Chromosomes replicate

    - Each duplicated chromosome consists of two identical sister chromatids and are attached at the centromeres

- Prophase I

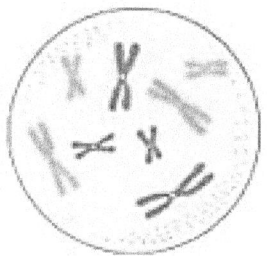

  - Chromosomes condense

  - Synapsis occurs; homologous chromosomes come together as pairs

  - Chromosomes condense further

    - Until each homologous pair in synapsis appears as a complex of four chromatids (tetrad)

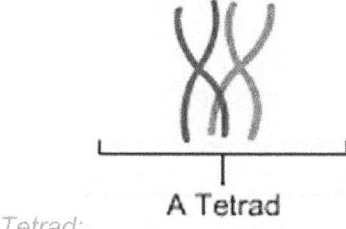

Tetrad: A Tetrad

  - In each tetrad, sister chromatids of the same chromosome are attached at their centromeres

    - Nonsister chromatids are linked by X-shaped chiasmata

    - Chiasmata - sites where homologous strand exchange or crossing-over occurs

- Chromosomes thicken more
  - Detach from the nuclear envelope
- Centriole pairs move apart
  - Spindle microtubules form between them
- Nuclear envelope and nucleoli disperse
- Chromosomes begin moving to the metaphase plate

- Metaphase I

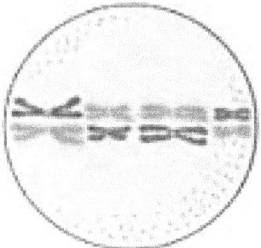

- Each synaptic pair is aligned so that centromeres of homologues point toward opposite poles
- Each homologue is attached to kinetochore microtubules emerging from the pole it faces

- Anaphase I

- Homologues separate and are moved toward the poles by the spindles
- Sister chromatids remain attached at their centromeres and move together toward the same pole
  - Homologue moves toward the opposite pole
  - Different from mitosis during which chromosomes line up individually on the metaphase plate and sister chromatids are moved apart toward opposite poles of the cell

- Telophase I and Cytokinesis

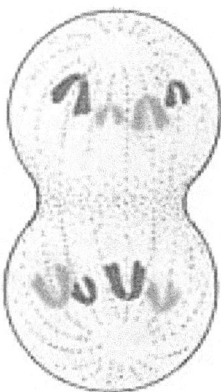

  o Each pole has a haploid set of chromosomes each composed of two sister chromatids attached at the centromere

  o Cytokinesis occurs simultaneously with telophase I forming two haploid daughter cells

  o In animal cells, a cleavage furrow forms

    ▪ In plant cells, a cell plate forms

  o In some species, nuclear membranes and nucleoli reappear, and the cell enters a period of interkinesis before meiosis II start

    ▪ In other species, the daughter cells immediately prepare for meiosis II

    ▪ In either case, no DNA replication occurs before meiosis II

Stages of Meiosis II

Meiosis II separates sister chromatids of each chromosome.

- Prophase II

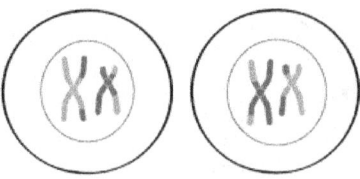

  o If the cell entered interkinesis

    ▪ Nuclear envelope and nucleoli disperse

  o Spindle apparatus forms and chromosomes move toward the metaphase II plate

- Metaphase II

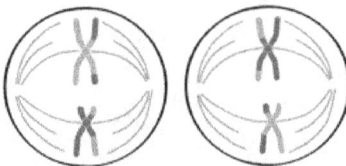

  - Chromosomes align singularly on the metaphase plate
  - Kinetochores of sister chromatids point toward opposite poles

- Anaphase II

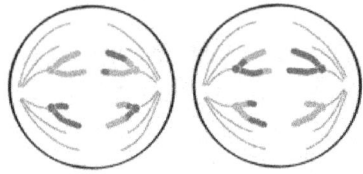

  - Centromeres of sister chromatids separate
  - Sister chromatids of each pair (now individual chromosomes) move toward opposite poles of the cell

- Telophase II and Cytokinesis

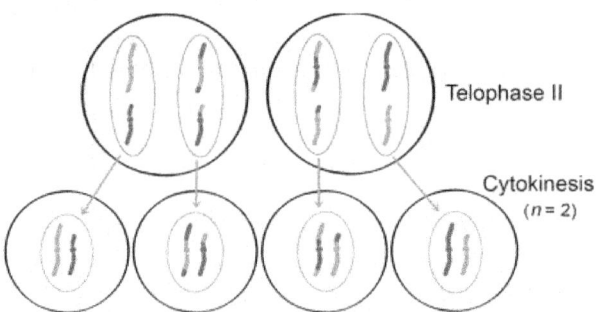

  - Nuclei form at opposite poles
  - Cytokinesis occurs producing four haploid daughter cells

Comparing Mitosis and Meiosis

- Mitosis produces two diploid daughter cells
  - Meiosis produces four haploid daughter cells
- Mitosis produces genetically identical daughter cells
  - Meiosis produces daughter cells that are not genetically identical

Spermatogenesis

Spermatogenesis refers to the production of sperm and it occurs in the testes of male animals.

- Diploid spermatogonium cell divides mitotically to produce two cells
    - One remains a spermatogonial cell
        - Other becomes a primary spermatocyte
- Primary spermatocyte progresses through meiosis I and meiosis II
    - Meiosis I yields two haploid secondary spermatocytes
    - Meiosis II yields four haploid spermatids
        - Each spermatid matures into a haploid sperm cell

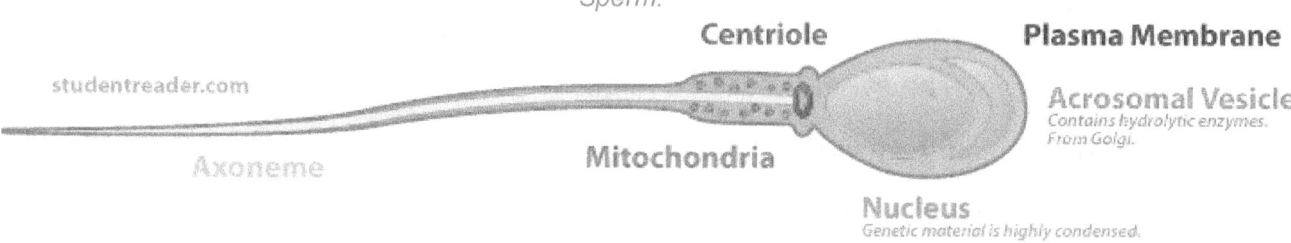

Sperm:

Centriole Plasma Membrane

studentreader.com

Acrosomal Vesicle
Contains hydrolytic enzymes.
From Golgi.

Axoneme Mitochondria

Nucleus
Genetic material is highly condensed.

- Sperm consist of a long flagellum and a head
    - Head contains a haploid nucleus
        - Capped by the acrosome
            - Acrosome contains digestive enzymes
            - Enzymes help sperm penetrate the protective layers of the egg

Oogenesis

Oogenesis refers to the production of egg cells and it occurs in the ovaries of female animals.

- Diploid oogonia produce diploid primary oocytes
    - Happens early in development

- Primary oocytes initiate meiosis I
    - But enter a dormant phase
        - Stuck in prophase I until the female becomes sexually mature
    - At puberty, primary oocytes are periodically activated to continue meiosis I
        - In humans one oocyte is activated per month
- Division in meiosis I is asymmetric
    - It produces two haploid cells of unequal size
        - A large secondary oocyte
        - A small polar body
- Secondary oocyte enters meiosis II but is held in that phase
    - It is released into the oviduct
        - The event is called ovulation
- If the secondary oocyte is fertilized
    - Meiosis II is completed
        - A haploid egg and a second polar body are produced
- The haploid egg and sperm nuclei then fuse to create the diploid nucleus of a new individual

Genetic Variation

Sexual reproduction contributes to genetic variation by independent assortment, crossing over during prophase I of meiosis, and the random fusion of gametes during fertilization.

Independent Assortment of Chromosomes

- At metaphase I, each homologous pair of chromosomes aligns on the metaphase plate
    - Each pair consists of one maternal and one paternal chromosome

- o Orientation of the homologous pair to the poles is random

    - Daughter cell of meiosis has a 50-50 chance of receiving either a maternal or paternal chromosome

    - o Gamete produced by meiosis contains just one of all the possible combinations of maternal and paternal chromosomes

- Independent assortment - random distribution of maternal and paternal homologues to the gametes

    - o Assortment may also refer to the random distribution of genes located on different chromosomes

- Each homologous pair assorts independently from all the others

    - o Process produces 2^n (n refers to the haploid number) possible combinations of maternal and paternal chromosomes in gametes

    - o In humans this is 2^{23}, about eight million possible combinations

Crossing Over

- Crossing over - exchange of genetic material between homologues

    - o Occurs during prophase of meiosis I

        - When homologous portions of two non-sister chromatids trade places

    - o X-shaped chiasmata becomes visible at places where this exchange occurs

    - o Produces chromosomes that contain genes from both parents

    - o In humans the rate is about two or three crossovers per chromosome pair

Random Fertilization

- In humans, an ovum represents one of eight million possible chromosome combinations

    - o Sperm cell represents another one of eight million possible combinations

- When fertilization occurs the resulting zygote can have one of 64 trillion possible combinations

    - o Without considering variations from crossing over

Chromosome Theory of Inheritance

Proposed at the turn of the 20th century, the chromosome theory of inheritance describes how the transmission of chromosomes accounts for the Mendelian patterns of inheritance.

- Developed from scientific inquiries:
    - Analysis of the transmission of traits from parent to offspring
    - Determination of the material basis of heredity
    - Examination of mitosis, meiosis, and fertilization
- Based on principles:
    - Chromosomes contain the genetic material
    - Chromosomes are replicated
        - Then passed from parent to offspring
    - Nuclei of most eukaryotic cells contain chromosomes
        - Found in homologous pairs
    - During meiosis, each homologue separates into one of the two daughter nuclei
    - When gametes are formed, non-homologous chromosomes segregate independently
    - Each parent contributes one set of chromosomes to its offspring
- How chromosomes help explain Mendelian patterns of inheritance
    - Mendel's law of segregations
        - Explained by homologous pairing and separation of chromosomes during meiosis
    - Mendel's law of independent assortment
        - Explained by the relative behavior of non-homologues chromosomes during meiosis

CHAPTER 5: EXTENSIONS OF MENDELIAN INHERITANCE

Extensions of Mendelian Inheritance

There are patterns of inheritance not described by Mendel, but his laws of segregation and independent assortment can be extended to these other cases.

- Mendelian inheritance patterns obey
    - Law of segregation
    - Law of independent assortment
- Mendelian inheritance involves
    - Single gene with two alleles
        - Alleles display simple dominant or recessive relationship
- There are traits that deviate from simple dominant/recessive relationship
    - Inheritance patterns of these traits still obey Mendelian laws

Alleles

Alleles are one of two or more alternative forms of a gene located on the same place on a chromosome.

- Wild-type alleles – the most prevalent alleles in a population
    - Typically encode proteins that function normally and are produced in the proper amount
- Mutant alleles – alleles altered by mutation
    - Rare in natural population
    - Likely cause reduction in the amount or function of the encoded protein
    - Often inherited in a recessive pattern
    - Often cause genetic diseases

- In simple dominant/recessive relationship, the phenotype of the heterozygote is not affected by the recessive allele

  - May be due to: ½ of the normal protein amount being sufficient to carry out the protein's cellular function

  - Or the heterozygote may produce more than ½ of the functional protein

    - Normal gene becomes up-regulated to compensate

- Essential genes – genes essential for survival

  - Absence of the protein encoded by the gene is lethal

- Nonessential genes – genes not necessary for survival

  - May confer benefits but not absolutely essential

- Lethal alleles – allele with the potential to cause the death of an organism

  - Can result from mutations in essential genes

  - Many, prevent cell division

    - Kills organism very early in development

  - Some lethal alleles exert their effect later in life

    - Example: Huntington disease age of onset is usually between 30 to 50

- Conditional lethal alleles – alleles that may cause the death of an organism if certain environmental conditions prevail

- Semi-lethal alleles – kill some individuals but not all individuals that are carriers

  - Environmental factors and other genes may shield the individual from effects of these alleles

Sex-Linked Traits

- Genes that are found on only one of the two types of sex chromosomes are called sex-linked

- Males have a single X chromosome

  - Called "hemizygous" for their X-linked genes

- Genes on the Y chromosome are called "holandric" genes

- X and Y chromosome have a short regions of homology at one end
  - Genes in this homologous region follow the inheritance pattern of a gene on an autosome

Reciprocal Crosses

Reciprocal crosses are crosses between different strains in which the sexes are reversed.

- Used to determine whether a trait is carried on a sex chromosome or an autosome
  - X-linked traits do not behave identically in reciprocal crosses
- Example:

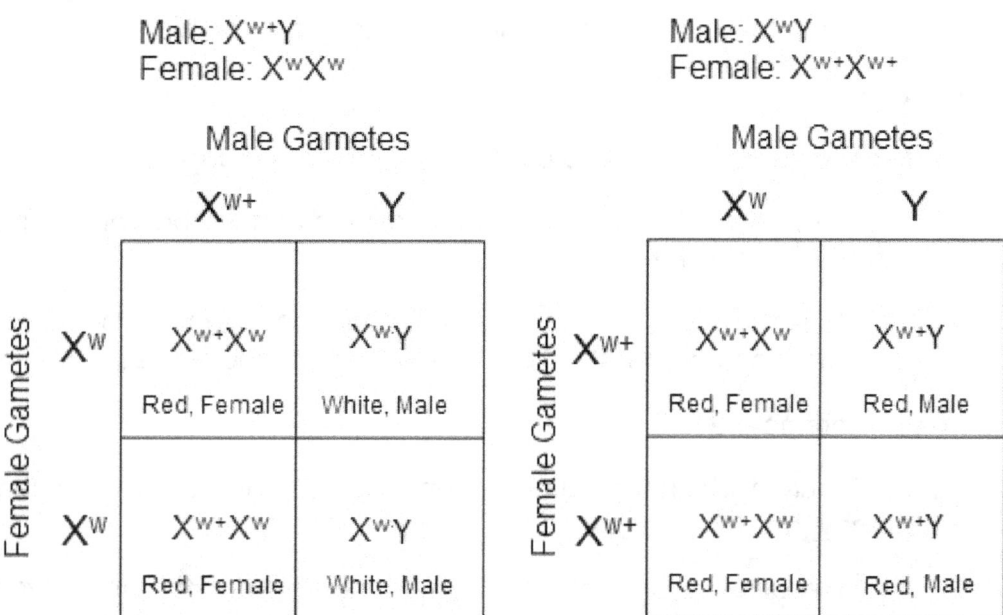

  - From the Punnett squares it becomes apparent that the reciprocal cross did not yield similar results
    - This is expected
      - Male transmit X-linked genes only to daughter
      - Female transmits X-linked genes to all children
    - This is why X-linked traits do not behave equally in reciprocal crosses

Different Types of Mendelian Inheritance Patterns

- Simple Mendelian

    - Inheritance of alleles that obey Mendel's laws and follow strict dominant/recessive relationship

        - Some genes can be found in three or more alleles

            - More complex relationship

- X-linked

    - Inheritance of genes located on the X chromosome

        - If a pair of x-linked alleles show a simple dominant/recessive relationship, 50% of the protein encoded by the dominant allele is enough to produce the dominant trait in a female

- Lethal alleles

    - Alleles with the potential to cause the death of an organism

    - Alleles are commonly loss of function alleles that encode proteins essential for survival

        - In rare cases, may be a nonessential gene that causes a protein to function abnormally with detrimental consequences

- Incomplete Dominance

    - Inheritance in which the dominant phenotype is not fully expressed in the heterozygote

        - Results in a phenotype intermediate between the homozygous dominant and homozygous recessive

        - 50% of the protein encoded by the wild-type (normal) allele is not sufficient to produce the normal trait

    - For example:

        - Red snapdragon (RR) is crossed with a white snapdragon (rr)

        - Resulting heterozygote progeny with Rr phenotype have pink flowers

        - Heterozygote produces half as much red pigment as the homozygous red-flowered plant

- Codominance
  - Inheritance characterized by full expression of both alleles in the heterozygote
  - Codominant alleles encode proteins that function slightly differently from each other
    - Function of each protein affects the phenotype uniquely
- Overdominance
  - Occurs when the heterozygote has a trait more beneficial than either homozygote
  - Potential benefits:
    - Cells are resistant to infection by certain microorganisms
    - Produces protein dimers with enhanced function
    - Produces proteins capable of functioning in a wider range of conditions
- Incomplete penetrance
  - Occurs when a dominant phenotype is not expressed despite an individual carrying a dominant allele
  - Protein encoded by the gene may not exert its effects
    - May be due to environmental influences
    - Or another gene may encode proteins that counteracts the effects of the protein encoded by the dominant allele
- Sex-influenced inheritance
  - Impact of gender on the phenotype of an individual
    - Some alleles are recessive in one sex and dominant in the other (e.g. baldness)
  - Sex hormones may regulate the expression of the genes

- Sex-limited inheritance

  - Traits that occur in only one of the genders

  - Sex hormones may regulate the expression of genes

    - Sex hormones primarily produced in only one gender are essential for producing a particular phenotype

- Multiple Alleles

  - Some genes may have more than just two alternative forms of a gene

  - ABO blood type is an example of a locus with three alleles

    - Paired combinations of three alleles produce four possible phenotypes:

      - Blood types: A, B, AB, or O

    - A and B are designations for genetically determined polysaccharides (A and B antigens) which are found on the surface of red blood cells

  - There are three alleles for this gene: I^A, I^B, and i

    - I^A allele codes for the production of A antigen

    - I^B allele codes for the production of B antigen

    - i allele codes for no antigen production

  - Every person carries only two alleles which specify their ABO blood type

  - From three alleles there are six possible genotypes

Blood Type	Possible Genotypes	Antigens on the Red Blood Cell	Antibodies in the Serum
A	$I^A I^A$ $I^A i$	A	anti-B
B	$I^B I^B$ $I^B i$	B	anti-A
AB	$I^A I^B$	A, B	----
O	ii	----	anti-A anti-B

ABO Blood Type, Possible Genotypes:

CHAPTER 6: NON-MENDELIAN INHERITANCE

Non-Mendelian Inheritance

Most genes in eukaryotes follow Mendelian patterns of inheritance but there are many that do not.

- Deviate from Mendelian Pattern because of:
    - Maternal effect and epigenetic inheritance
        - Involve genes in the nucleus
    - Extranuclear inheritance
        - Involve genes in organelles beside the nucleus
            - Mitochondria
            - Chloroplasts

Maternal Effect

Maternal effect refers to the inheritance pattern in which the genotype of the mother in specific nuclear genes determines the phenotype of her offspring.

- Phenotype of each generation depends on the maternal genotype of the generation preceding it
    - Phenotype of the progeny is determine by the mother's **genotype**
        - Not the mother's **phenotype**
    - Genotypes of both the father and the offspring have no effect on the phenotype of the offspring
- Maternal effect genes encode RNA or proteins that are essential in early development

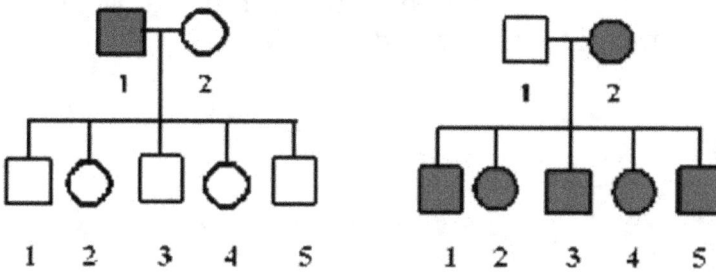

Maternal Inheritance Pedigrees:

Epigenetic Inheritance

Epigenetic inheritance refers to a pattern of inheritance in which a modification occurs to a nuclear gene or chromosome that alters gene expression.

- Expression not affected over the course of many generations

- Epigenetic changes occur from modifications made to DNA or chromosomes

  - Occur during oogenesis, spermatogenesis, or early embryonic development

Dosage Compensation

Dosage compensation refers to the equalization of gene expression between males and females of a species.

- Occurs because different sex chromosomes contain different numbers of genes

- Dosage compensation occurs via different mechanisms depending on the species

  - Females have two X chromosomes while males have a single X chromosome

  - One of the X chromosomes in the somatic cells of females is inactivated

    - In some species the paternal X chromosome is always the one inactivated

    - Species such as humans, however, either the maternal or paternal X chromosome is randomly inactivated

X chromosome Inactivation Mechanism

Mechanism of X chromosome inactivation is also called the Lyon hypothesis.

- X chromosome that is inactivated becomes highly compacted

  - Most genes are not accessible and can't be expressed

- When the inactivated chromosome is replicated during cell division, both copies remain compacted and inactive

Genomic Imprinting

Genomic imprinting refers to the phenomenon in which expression of a gene depends on whether the gene is inherited from the father or the mother.

- Imprinting is permanent in somatic cells of a given generation
- Imprinting may involve
    - A single gene
    - Part of a chromosome
    - Entire chromosome
    - Or all chromosomes

Methylation

- Genes are "marked" in such a way that offspring expresses either the maternal or paternal allele
    - Aka monoallelic expression
    - Marking of alleles can be altered from generation to generation
- Marking in some genes is known to involve differentially methylated regions (DMRs)
    - Located near the imprinted genes
    - Methylated in either the oocyte or sperm
        - Not both
    - Contain binding sites for one or more transcriptional factors
- In most genes methylation causes inhibition of gene expression

Extranuclear Inheritance

Extranuclear inheritance refers to inheritance patterns involving genetic material outside of the nucleus.

- Mitochondria and chloroplasts are most important examples
    - Both organelles have genetic material in a region called the nucleoid
        - Genome is a single circular chromosome consisting of double-stranded DNA

- Mitochondrial genomes are small in animals

  o Large in plants

- Genetic material of mitochondria is called mtDNA

  o In human, carries 13 genes that are needed for oxidative phosphorylation

    ▪ Oxidative phosphorylation is a process of generating ATP

  o Most mitochondrial proteins are encoded by genes in the nucleus

  o Human mtDNA is transmitted from mother to offspring, not from father to offspring for several reasons:

    ▪ Egg has more mitochondria than a sperm cell

    ▪ Egg marks mitochondria from the sperm for destruction

- Genetic material of chloroplasts is called cpDNA

  o 10 times larger than mitochondrial genome in animals

  o Many chloroplast proteins are encoded by genes in the nucleus

Endosymbiotic Theory

Endosymbiotic theory describes the evolutionary origin of mitochondria and chloroplasts. Large bacterial cells lost their cell walls and engulfed smaller bacteria and a symbiotic (mutualistic) relationship developed between them. The host cell supplied the nutrients, and the engulfed cell produced excess energy that the host could use. Evidence for the theory:

- Mitochondria and chloroplasts resemble bacteria in size and shape

- They divide on their own - independent of the host and the process is nearly identical to binary fission

- They contain their own DNA

  o A single circular chromosome

- They contain 70S ribosomes

Endosymbiotic Theory

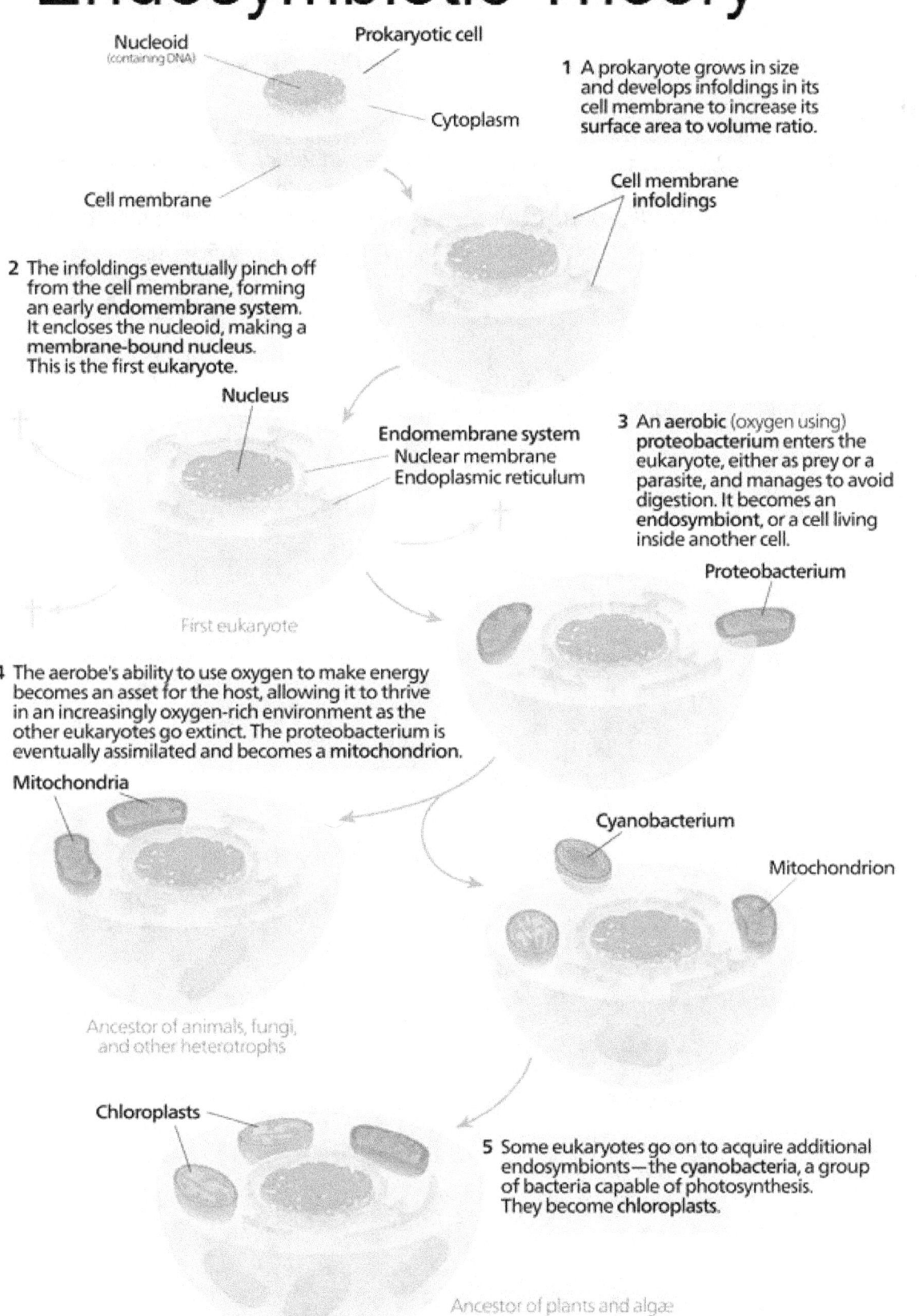

Nucleoid
(containing DNA)

Prokaryotic cell

Cytoplasm

Cell membrane

1 A prokaryote grows in size and develops infoldings in its cell membrane to increase its surface area to volume ratio.

Cell membrane infoldings

2 The infoldings eventually pinch off from the cell membrane, forming an early **endomembrane system**. It encloses the nucleoid, making a membrane-bound **nucleus**. This is the first **eukaryote**.

Nucleus

Endomembrane system
Nuclear membrane
Endoplasmic reticulum

3 An **aerobic** (oxygen using) **proteobacterium** enters the eukaryote, either as prey or a parasite, and manages to avoid digestion. It becomes an **endosymbiont**, or a cell living inside another cell.

Proteobacterium

First eukaryote

4 The aerobe's ability to use oxygen to make energy becomes an asset for the host, allowing it to thrive in an increasingly oxygen-rich environment as the other eukaryotes go extinct. The proteobacterium is eventually assimilated and becomes a **mitochondrion**.

Mitochondria

Cyanobacterium

Mitochondrion

Ancestor of animals, fungi, and other heterotrophs

Chloroplasts

5 Some eukaryotes go on to acquire additional endosymbionts—the **cyanobacteria**, a group of bacteria capable of photosynthesis. They become **chloroplasts**.

Ancestor of plants and algæ

Other Non-Mendelian Patterns of Inheritance

Pleiotropy

Pleiotropy is the ability of a single gene to have multiple phenotypic effects.

Epistasis

Epistasis is the interaction between two non-allelic genes in which one modifies the phenotypic expression of the other (i.e. a gene at one locus alters the phenotypic expression of a second gene). If a gene suppresses the phenotypic expression of another gene, the suppressing gene is said to be epistatic to the gene being suppressed.

Polygenic Inheritance

Polygenic inheritance is a mode of inheritance in which the additive effect of two or more genes determines a single phenotypic character.

CHAPTER 7: GENETIC LINKAGE AND CHI-SQUARE ANALYSIS

Linkage and Crossing Over

- In eukaryotes, typical chromosome contains many hundred or even a few thousand genes

- Linkage – tendency of alleles located close together on a chromosome to be inherited together during meiosis

- Linkage group – set of genes at different loci on the same chromosome that tend to act as a single pair of genes in meiosis instead of undergoing independent assortment

- Genes far apart on the same chromosome may assort independently from each other

- Linkage can be altered during meiosis as a result of crossing over

    o Crossing over occurs during prophase I of meiosis

    o Non-sister chromatids of homologous chromosomes exchange DNA segments

Chi Square Analysis

Chi square analysis is a method used to determine whether the outcome of a cross is consistent with linkage or independent assortment.

- Chi-square test is a "goodness of fit" test

    o Meant to answer the question of how well a set of experimental data fits with expectations

- A chi-square test begins with a null hypothesis

    o Usually, the null hypothesis is the predictive ratio of a cross

        ▪ Based on the phenotype of the observed and basic gene theory

- Next, calculate chi-square using the formula:

    o $$X^2 = \sum \frac{(observed - expected)^2}{expected}$$

- o X is the Greek letter chi

- o Σ is sigma and means that you have to sum everything that appears after it

- o Observed is the number of individuals of the given phenotype that is observed

  - ▪ Usually this value is given in the question

- o Expected is the number of individuals of the given phenotype expected from the null hypothesis

  - ▪ Calculated by multiplying the total offspring number by the expected proportions

- Next, figure out the degrees of freedom

  - o Degrees of freedom is the number of outcome classes minus one

- Next, look up critical values for chi-square on a table that is sorted by degrees of freedom and probability values (look at the table that follows as an example)

  - o Most commonly the $p = 0.05$ is used (your instructor may tell you to use another value)

  - o Find the degree of freedom you are working with on the leftmost row and follow the row horizontally to find the $p = 0.05$ row and the critical chi-square value

  - o If you get a chi-square value that is less than the critical value than you say that you "fail to reject" the null hypothesis

  - o If you get a chi-square value that is more than the critical value than you say that you "reject" the null hypothesis

df	0.995	0.975	0.9	0.5	0.1	0.05	0.025	0.01	0.005	df
1	.000	.000	0.016	0.455	2.706	3.841	5.024	6.635	7.879	1
2	0.010	0.051	0.211	1.386	4.605	5.991	7.378	9.210	10.597	2
3	0.072	0.216	0.584	2.366	6.251	7.815	9.348	11.345	12.838	3
4	0.207	0.484	1.064	3.357	7.779	9.488	11.143	13.277	14.860	4
5	0.412	0.831	1.610	4.351	9.236	11.070	12.832	15.086	16.750	5
6	0.676	1.237	2.204	5.348	10.645	12.592	14.449	16.812	18.548	6
7	0.989	1.690	2.833	6.346	12.017	14.067	16.013	18.475	20.278	7
8	1.344	2.180	3.490	7.344	13.362	15.507	17.535	20.090	21.955	8
9	1.735	2.700	4.168	8.343	14.684	16.919	19.023	21.666	23.589	9
10	2.156	3.247	4.865	9.342	15.987	18.307	20.483	23.209	25.188	10
11	2.603	3.816	5.578	10.341	17.275	19.675	21.920	24.725	26.757	11
12	3.074	4.404	6.304	11.340	18.549	21.026	23.337	26.217	28.300	12
13	3.565	5.009	7.042	12.340	19.812	22.362	24.736	27.688	29.819	13
14	4.075	5.629	7.790	13.339	21.064	23.685	26.119	29.141	31.319	14
15	4.601	6.262	8.547	14.339	22.307	24.996	27.488	30.578	32.801	15

Example of Applying Chi-square Analysis

- You count the F_1 offspring of a cross and find that 290 plants had purple flowers and 110 white flowers

    o Total of 400 (290 + 110) offspring

    o Let's say that in the null hypothesis you predicted a 3:1 ratio

    o Next calculate the expected number of offspring

        ▪ This is done by multiplying the total offspring number by the expected proportions

        ▪ 400 * 3/4 = 300 plants with purple flowers

        ▪ 400 * 1/4 = 100 plants with white flowers

    o Now you can just plug numbers into the formula

    o $X^2 = \frac{(290-300)^2}{300} + \frac{(110-100)^2}{100} = 1.333..$

    o Find the degrees of freedom you are working with

        ▪ There are 2 outcome classes (purple-flowered and white-flowered plant)

        ▪ Degrees of freedom is the number of outcome classes minus 1 so:

            • 2 - 1 = 1

        ▪ The degrees of freedom is 1

- The critical chi-square value for degree of freedom of one and a p value of 0.05 is 3.841 as determined from the chart

  - Since the calculated chi-square value of 1.333... is less than the critical value of 3.841 you say that you "fail to reject" the null hypothesis

CHAPTER 8: BACTERIAL GENE TRANSFER

Introduction to Bacterial Genetics

- Usually, bacteria are haploid

    o Makes it easier to study loss-of-function mutations

- Bacteria reproduce asexually

    o Offspring is a genetic clone of the parent

        ▪ Barring any mutations

- Bacterial genetics focuses heavily on genetic transfer

- Transfer of genetic material in bacteria can occur in three ways:

    o Conjugation

        ▪ Involves direct physical contact

    o Transduction

        ▪ Involves viruses

    o Transformation

        ▪ Involves uptake from the environment

Genetic Transfer and Recombination

Genetic recombination

- Exchange of genes between two DNA molecules to form new combinations of genes on a chromosome

    o Crossing over in eukaryotes

- Increases a population's genetic diversity

- Recombination is more likely to produce a beneficial outcome than mutation

    o Less likely to destroy a gene's function

    o May bring together a combination of genes that enable the organism to carry out a valuable new function

Vertical Gene Transfer

- Genes are passed from parent cell to a daughter cell

Horizontal Gene Transfer

- Genes are passed from one adult cell to another

- Donor cell gives a portion of its DNA to a recipient cell

    o Part of the donor's DNA is incorporated into the recipient's DNA

        ▪ Recipient is then called a recombinant cell

- Very rare occurrence

    o May happen in less than 1% of a population

Conjugation in Bacteria

Bacterial conjugation refers to the transfer of genetic material between bacterial cells by direct cell-cell contact or a bridge-like connection between the two cells.

- Only certain strains of bacteria can donate the genes they carry

    o Contain a plasmid carrying the F factor

    o Plasmid itself is called the F plasmid

        ▪ Plasmid - small, circular piece of DNA that replicates independently from the cell's chromosome

            • Plasmids transferred during conjugation are called conjugative plasmids

        ▪ Strains that have the F factor are referred to as being F^+

            • Cells lacking it are F^-

- In F+ cells, plasmids carry genes that code for the synthesis of a sex pili

    o Sex pili help bring two cells together for the transfer

- Bacterial conjugation steps
  - Conjugation is initiated between a F+ and F- cell
  - One strand of the F factor is cut by an endonuclease
    - Strand moves across the conjugation tube
  - DNA complement is synthesized on both single strands
  - Ligase seals both strands reforming the plasmid in both the cells
  - Conjugates separate
    - Both cells are F$^+$ at the end since they both contain the F plasmid

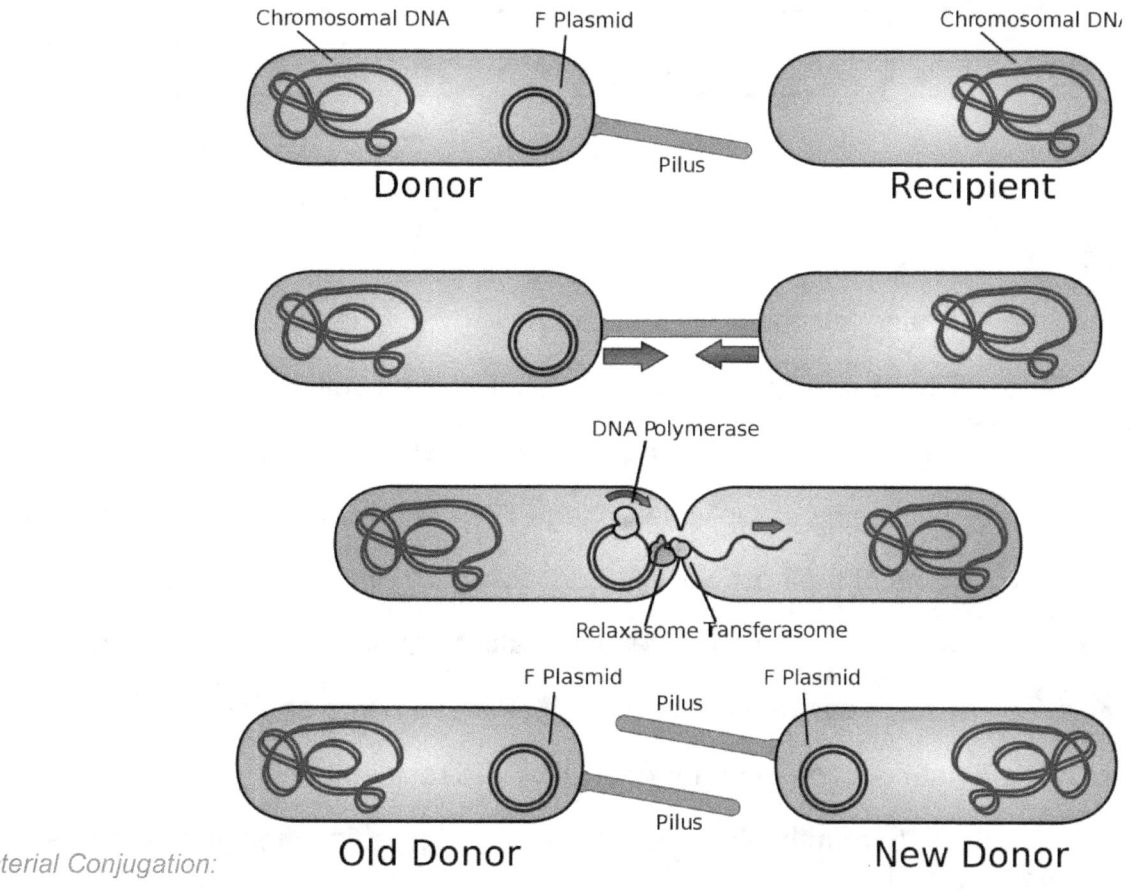

Bacterial Conjugation:

Transduction in Bacteria

Transfer of genetic material using a genetic vector. The vector is a bacteriophage or simply a phage, which are viruses that infect bacteria.

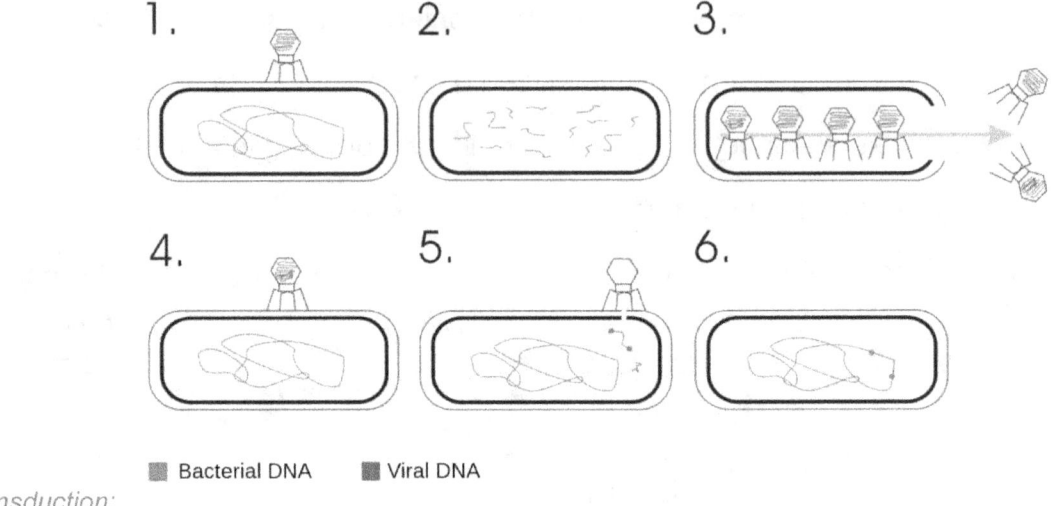

■ Bacterial DNA   ■ Viral DNA

Transduction:

Steps

- Step 1: Phage latches on to a bacterium

- Step 2: Phage injects its genetic material into the cell

  - Phage DNA replication occurs and host genome is fragmented

- Step 3: Packaging of replicated phage DNA and synthesis of new phages

  - Some of the host cell DNA is packaged with the viral DNA

  - Cell lyses open and releases mature phage into its surroundings

- Step 4: Phage carrying donor host DNA latches onto another cell

- Step 5: Introduction of donor host DNA into recipient cell

- Step 6: DNA from the first bacterial host is recombined into the chromosome of the new host

Bacteriophage Multiplication

Bacteriophages are DNA viruses that multiply by two alternative mechanisms: the lytic cycle and the lysogenic cycle.

- Lytic Cycle
    - Phage attaches by tail fibers to the host cell
        - Viral attachment site binds to a complimentary receptor site (protein/antigen) on the bacterium
    - Phage lysozyme opens the cell wall and its tail sheath contracts to insert the tail core and viral DNA into the cell
    - Phage DNA circularizes and directs production of phage DNA and proteins
        - Host DNA is fragmented
    - Maturation - spontaneous assembly of phage components into virions
    - Phage lysozyme breaks cell wall and lyses open the host cell to release new virions
- Lysogenic Cycle
    - Lysogenic phages may proceed through a lytic cycle, but they can also insert their DNA into the host cell's DNA and begin a lysogenic cycle
    - In lysogeny, the phage remains inactive
        - Host cell remains functional during lysogeny and is called a lysogenic cell
    - Phage DNA recombines with host DNA to form a prophage
        - Most prophage genes are repressed by repressor proteins
        - Lysogenic bacterium reproduces normally
    - Occasionally, the prophage may excise itself from the bacterial chromosome to initiate the lytic cycle

Transformation in Bacteria

Transformation is the process by which a bacterium takes up extracellular DNA.

- When a cell lyses, its DNA is released

- Closely related living cells can take up fragments of that DNA and incorporate it into their own DNA

- Forms a hybrid recombinant cell

- All descendants of the recombinant will be identical to it

- Transformation naturally occurs only among a few genera of bacteria

CHAPTER 9: VARIATION IN CHROMOSOME STRUCTURE AND NUMBER

Genetic Variation

Genetic variation can refer to genetic differences between individuals within the same species or genetic variation between different species.

- Allele variation occurs from mutations in particular genes

- Chromosomal mutations refer to substantial changes in chromosome structure

    o Usually, affects more than one gene

- Genome mutation – change in chromosome number

    o May refer to change in either the number of sets of chromosomes or the number of individual chromosomes in a set

- Chromosomal mutations have major effects on the phenotype of an organism

    o Or its offspring

Cytogenetics

Cytogenetics is a branch of genetics that focuses on the study of inheritance in relation to the structure and function of chromosomes.

- Chromosomal composition of a particular cell or organism is examined

    o Detects abnormal chromosome number or structure

    o Also a way of distinguishing differences between species

- Main features of chromosomes that are identified and classified

    o Size

    o Location of centromere

    o Banding patterns

- Based on these differences, chromosomes from a photomicrograph can be matched into homologous pairs and arranged in a standard sequence to produce a karyotype

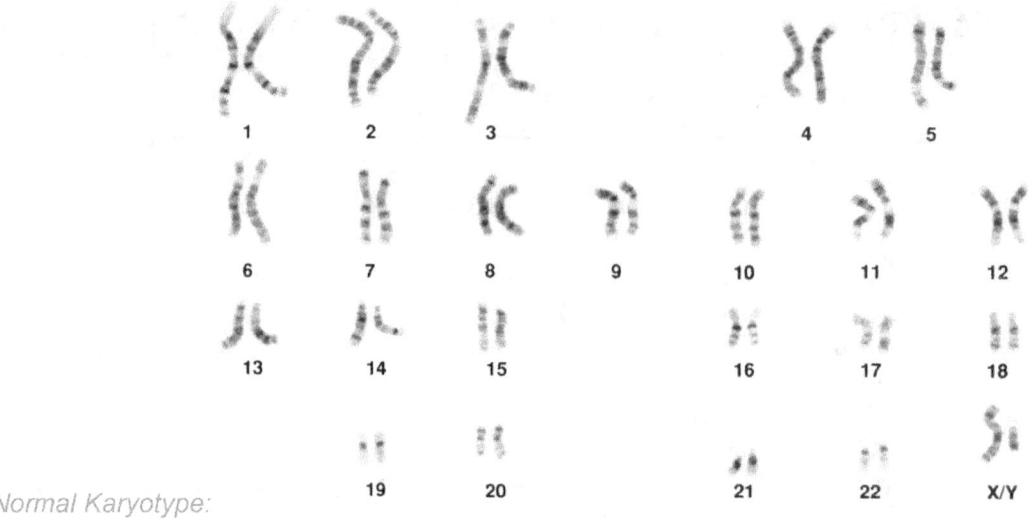

Normal Karyotype:

- Karyotype - display or photomicrograph of an individual's somatic-cell metaphase chromosomes that are arranged in a standardized sequence

  o Can be used to screen for chromosomal abnormalities

Chromosome Banding

- For detailed identification chromosomes are treated with stain to produce characteristic banding patterns

  o Example of G-banding

    ▪ Chromosomes are treated with Giemsa (a dye)

    ▪ Some regions bind to the dye heavily (produce dark bands)

    ▪ Some regions do not bind to the stain very well (produce light bands)

- Banding pattern helps with:

  o Distinguishing chromosomes from one another

  o Detecting changes in chromosome structure

  o Determining evolutionary relationships between chromosomes of different species

Altered Chromosome Structure

Altered chromosome structure can result from:

- Changes in the total amount of genetic information in the chromosome
    - Deletion/deficiency - loss of a chromosomal segment
    - Duplication - having an extra chromosomal segment
- Changes in the arrangement of genetic material
    - Inversions - change of direction of genetic material along a chromosome
        - Pericentric – inversion involving the centromere
        - Paracentric – inversion that does not involve the centromere
    - Translocations - segment of one chromosome is attached to another chromosome
        - Simple translocation – one way transfer
        - Reciprocal translocation – two way transfer (swap)

Altered Chromosome Number

- Euploidy – normal variation of the number of complete sets of chromosomes
    - Euploid number varies among species
- Aneuploidy – variation in the number of particular chromosomes within a set
    - Monosomy – lack a copy of a chromosome
    - Trisomy – extra copy of a chromosome
    - Polysomy – having one or more chromosomes than the normal amount
        - Generally, lethal in animals

Sex Chromosomal Aneuploidy

- Klinefelter Syndrome (XXY)
    - Characteristics: sexual immaturity (no sperm production), breast swelling
- Turner Syndrome (X)
    - Characteristics: short, webbed neck, sexually undeveloped

- Jacobs Syndrome (XYY)
  - Characteristics: tall
- Triple X Syndrome (XXX)
  - Characteristics: tall, thin, menstrual irregularity

Autosomal Aneuploidy

- Trisomy 21 (aka Down Syndrome)
  - Characteristics: mental retardation, slanted eyes, flattened face, short
- Trisomy 18 (aka Edward Syndrome)
  - Characteristics: mental and physical retardation, facial abnormalities, extreme musculature, early death
- Trisomy 13 (aka Patau Syndrome)
  - Characteristics: mental and physical retardation, defective organs, large nose, early death

CHAPTER 10: DNA AND RNA

Identifying DNA as the Genetic Material

We didn't always know that DNA was the genetic material. Identifying DNA as the genetic material involved many experimental approaches.

- Genetic material must have:
  - Information
    - Must contain information to make an entire organism
  - Transmission
    - Must be able to pass from parent to offspring
  - Replication
    - Must be able to be copied so it can be passed on
  - Variation
    - Must have the capacity for change to account of the variations seen between species and individuals within a species

The Griffith Experiment

- Griffith studied the two strains of the bacterium *Streptococcus pneumonia*
  - S – smooth strain
    - Has a polysaccharide capsule that produces smooth looking colonies when grown in a lab
    - Pathogenic strain
  - R – rough strain
    - No capsule; colonies appear rough
    - Nonpathogenic strain
- He injected a live smooth strain into a mouse
  - Mouse died
- He injected heat-killed smooth stain into a mouse
  - Mouse not affected

- He injected a live rough strain into a mouse

  o Mouse not affected

- He injected heat-killed smooth strain with live rough strain into a mouse

  o Mouse died

    ▪ Concluded that something transferred between the strain to allow the rough strain to kill the mouse

    ▪ Griffith called this process transformation

    ▪ Transformation – genetic alteration of a cell resulting from direct uptake and incorporation of exogenous genetic material

- Griffith didn't know what material allowed for transformation to occur

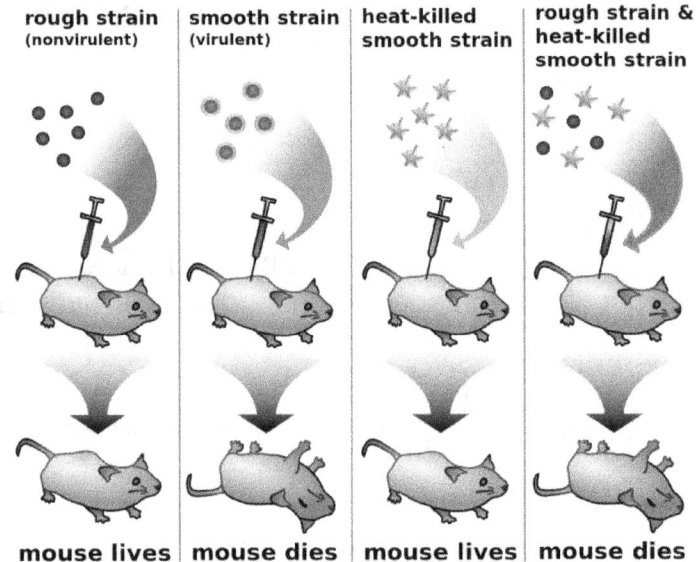

Griffith Experiment: **mouse lives** | **mouse dies** | **mouse lives** | **mouse dies**

The Avery, MacLeod, and McCarthy Experiment

- Knew that cells had large amounts of DNA, RNA, proteins, and carbohydrates that constituted them

  o All four were considered potential candidates for being the genetic material

- Knew Griffith's observations could be used to help identify genetic material

  o Attempted to transform the R strain by incubating them with heat-killed S strain

  o They pretreated the heat-killed S strain in four different ways in order to purify each type of macromolecule (DNA, RNA, proteins, and carbohydrates)

    ▪ Only the heat-killed S strain that contained the purified DNA was able to transform the R strain into the virulent strain

  o To verify DNA and not another contaminant (e.g. protein) was the transforming agent they did the following:

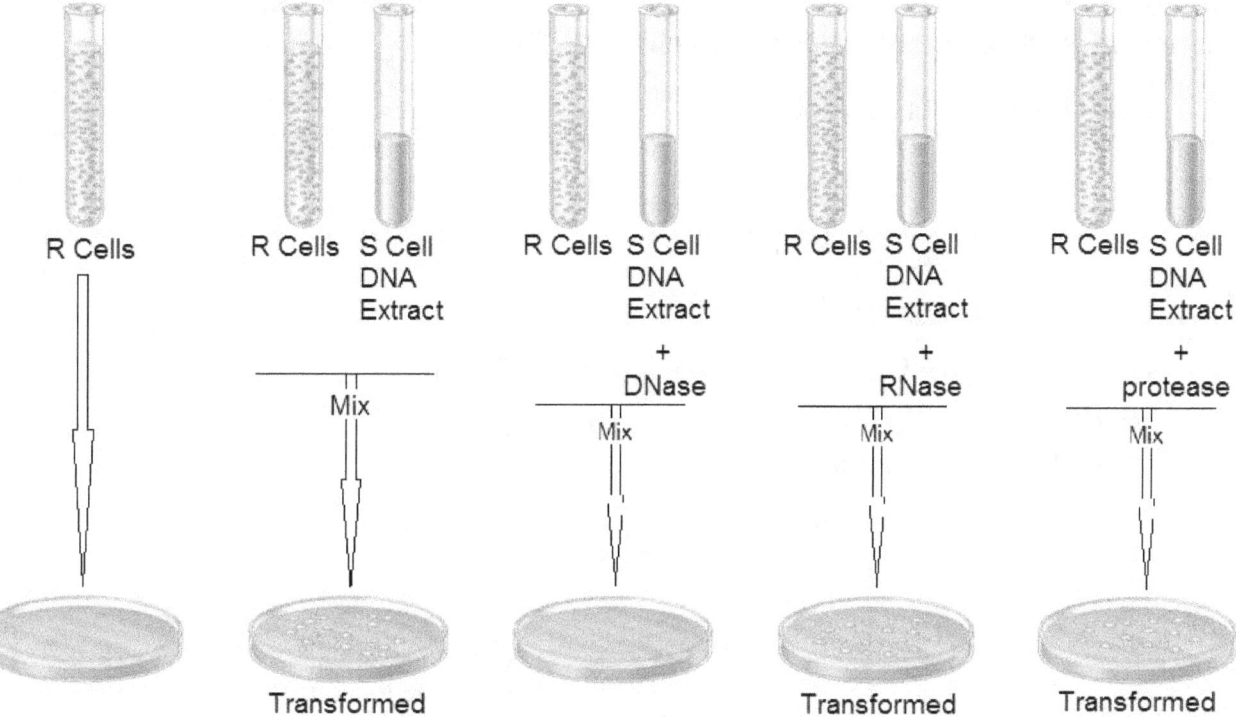

    ▪ Note: no transformation occurred where the S DNA was treated with DNase (degrades DNA), transformation occurred wherever S DNA was present

  o Showed that DNA is the transforming agent

    ▪ However, people were still skeptical of this result

Hershey and Chase Experiment

- Provided further evidence for DNA being the genetic material

- They studied bacteriophage T2

- Composed of only two macromolecules (DNA and protein)

- Used two different radioisotopes to tag DNA and proteins in the phages to distinguish them from one another

- Allowed the tagged phages to infect cells

    - Only the genetic material of the phage is injected into the bacterium

        - Isotope labeling showed that it was DNA and not protein that was injected into the bacterium

        - Proved DNA was the genetic material

Nucleotides and Nucleic Acids

Nucleotides (Monomers of Nucleic Acids)

- 3 components of nucleotides:

    - 5-carbon sugar (pentose)

    - Nitrogen containing base ring

    - PO_3 group (phosphate group)

        - Attached to the #5 carbon of the sugar

- Pyrimidine – nitrogenous base with a characteristic six-membered ring consisting of carbon and nitrogen atoms

    - Cytosine (C), Thymine (T), and Uracil (U) are pyrimidines

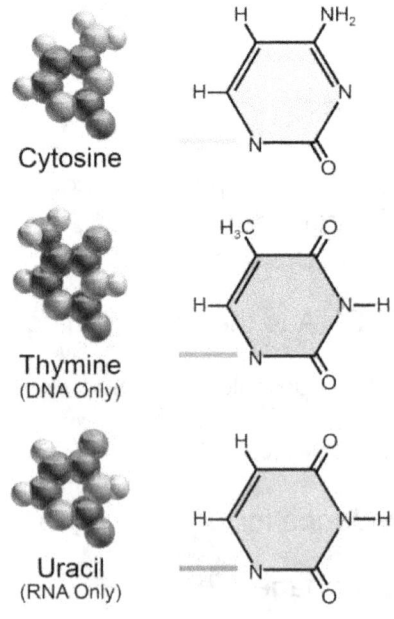

Cytosine

Thymine
(DNA Only)

Uracil
(RNA Only)

Pyrimidines:

- Purine – nitrogenous base with a characteristic five-membered ring fused to a six-membered ring

    o Adenine (A), and Guanine (G) are purines

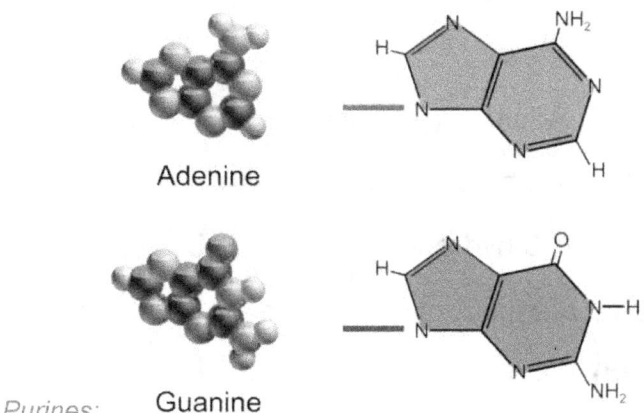

Adenine

Guanine

Purines:

- Phosphodiester bonds covalently link nucleotides together

    o Phosphate connects the 5' carbon of one nucleotide to the 3' carbon of another nucleotide

        ▪ Gives strands directionality

        ▪ In a strand, all sugars are oriented in the same direction

- Phosphates and sugar form the backbone of a strand

    o Bases project from the backbone

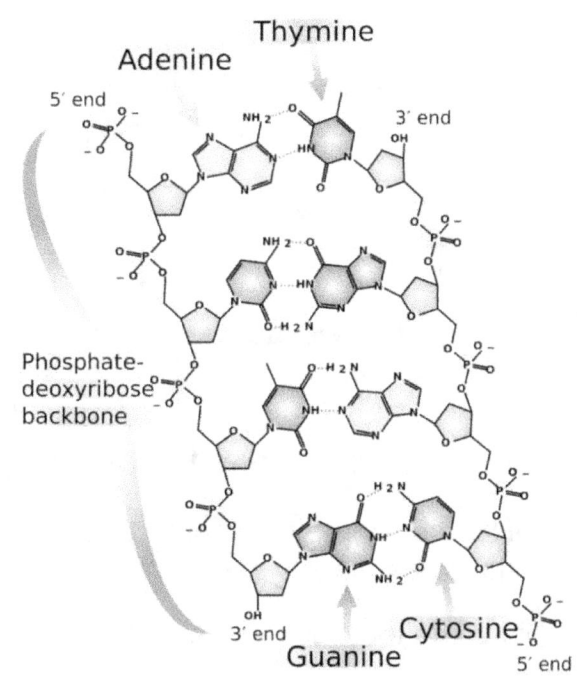

DNA Structure:

DNA (Deoxyribonucleic Acid)

- Deoxyribose as the pentose sugar

- Two nucleotide chains form a double helix

- Contains the nucleotides:

    o Thymine (T), Adenine (A), Cytosine (C), and Guanine (G)

        ▪ A forms 2 hydrogen bonds with T

        ▪ G forms 3 hydrogen bonds with C

- Contain genes that are instructions for protein synthesis

RNA (Ribonucleic Acid)

- Single stranded nucleic acid

- Ribose is the pentose sugar

- Contains same nucleotides as DNA except that Adenine (A) is replaced with Uracil (U)

- Functions in the synthesis of proteins

    o Messenger RNA (mRNA) carries encoded genetic message to the cytoplasm from the nucleus

RNA vs. DNA

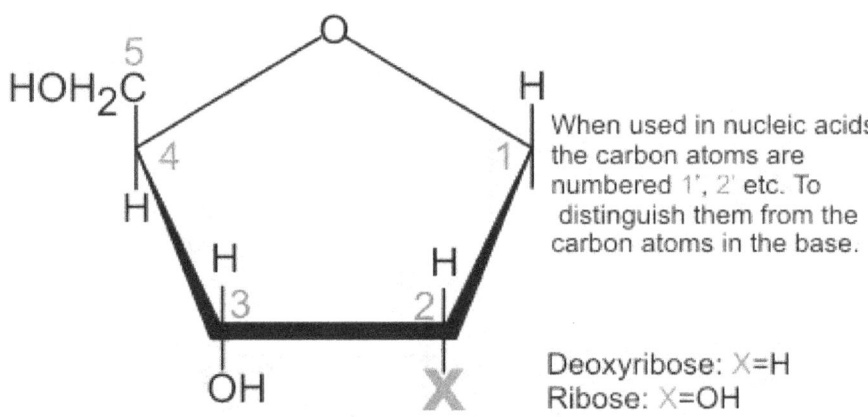

Deoxyribose & Ribose Sugars

- Position of 2'-OH of ribose prevents formation of classic Watson-Crick B helix in RNA due to steric hindrance

- 2'-O atom would come too close to 3 atoms of the adjoining phosphate & 1 atom in next base

Nucleoside vs. Nucleotide

- Base + sugar are nucleosides
    - Examples:
        - Adenine + ribose = adenosine
        - Adenine + deoxyribose = deoxyadenosine
- Base + sugar + phosphate(s) are nucleotides
    - Examples:
        - Adenosine monophosphate (AMP)
        - Adenosine diphosphate (ADP)
        - Adenosine triphosphate (ATP)

Chargaff's Rule

Chargaff analyzed base composition of DNA isolated from several species.

- He observed that consistently
    - Percent of adenine = percent of thymine
    - Percent of cytosine = percent of guanine
- His observations became known as Chargaff's rule
    - Rule states that DNA of all organisms should have a 1:1 ratio of pyrimidine and purine
        - More specifically, amount of guanine = amount of cytosine
            - Amount of adenine = amount of thymine

DNA Structure

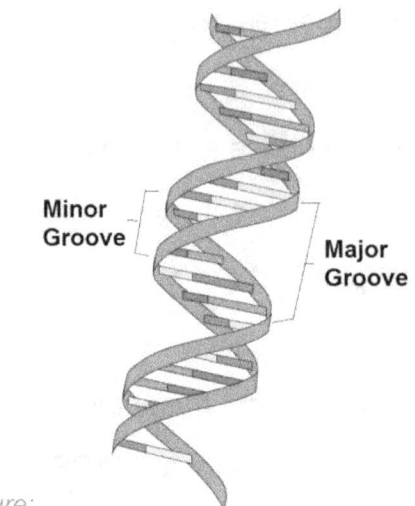

DNA Structure:

- Two strands are twisted together around a common axis

  - Strands are antiparallel

    - One runs 5' to 3'

    - Other runs 3' to 5'

- Helix is right-handed

  - Helix turns in a clockwise direction

  - If you were to grab the helix with your right hand and wrap your fingers along the minor groove

    - Following the minor groove upward, your hand follows your thumb

  - If you do the same with your left hand, your hand moves away from the direction your thumb points

- Helix is stabilized by

  - Hydrogen bonding between complementary bases

- Major groove vs. Minor groove

  - These grooves are spaces between the backbones

  - The major groove occurs where the backbones are far apart

  - The minor groove occurs where they are close together

RNA Structure

Primary structure of an RNA strand is very similar to that of DNA strand.

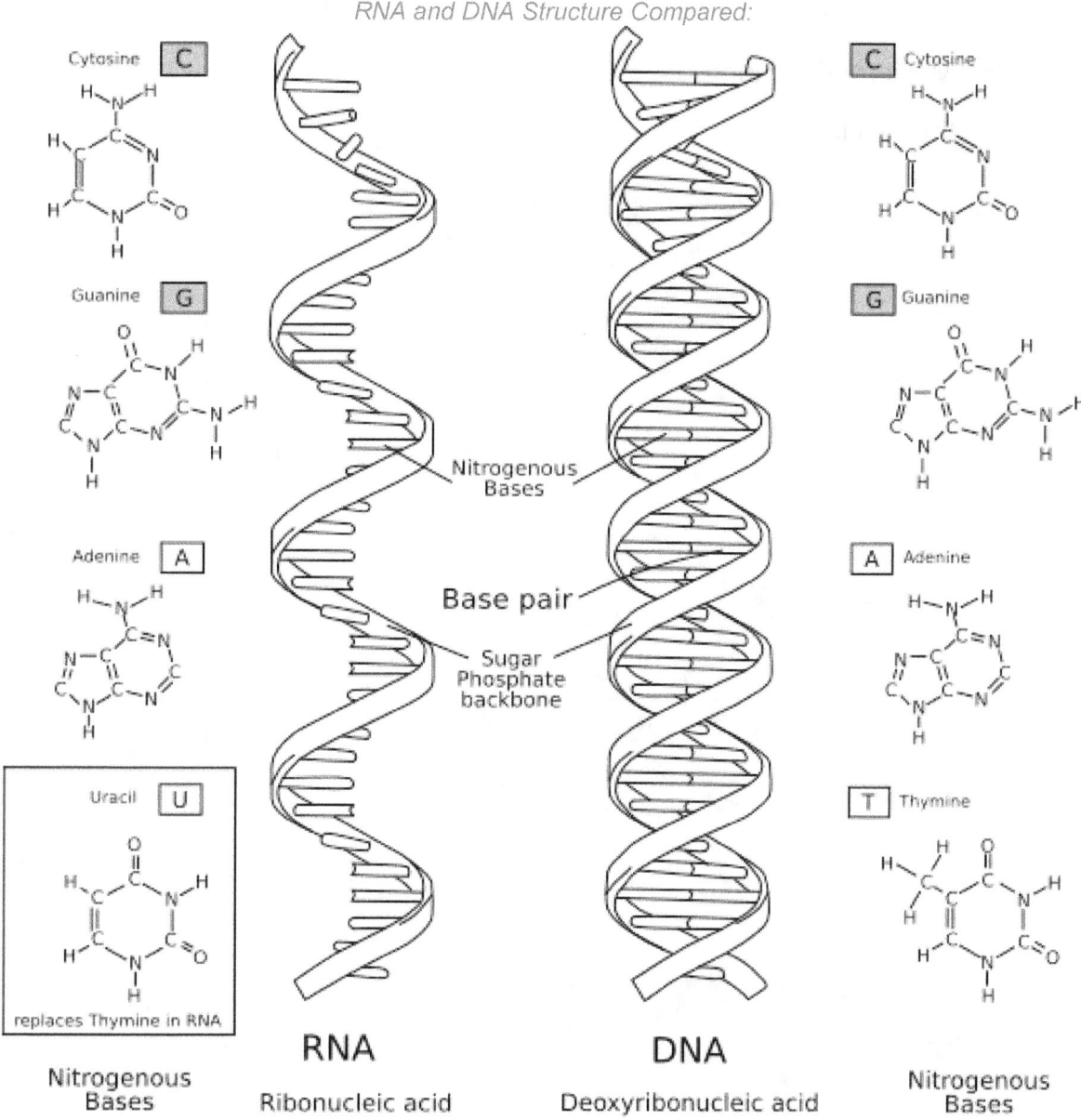

RNA and DNA Structure Compared:

- Only one strand in DNA is used as a template for RNA synthesis
- RNA is usually single stranded, but can form short double-stranded regions
  - Happens because of complementary base-pairing

- RNA double helices are typically right handed

- Different types of RNA secondary structures are possible

    o Stem loop

    o Multibranched junction

    o Internal loop

    o Bulge loop

CHAPTER 11: CHROMOSOME ORGANIZATION AND MOLECULAR STRUCTURE

Viral Genomes

Viral genome is infectious particles containing nucleic acid surrounded by a protein capsid.

- Genome can be

    o DNA or RNA

    o Single or double stranded

    o Circular or linear

    o Few thousand to more than hundred thousand nucleotides in size

- Rely on host cell for replication

    o Ribosomes, tRNA, amino acids, ATP, and most enzymes needed for protein and nucleic acid synthesis are supplied by the host

- Viruses have limited host range

    o May only affect specific cells in a specific species

Bacterial Genomes

- Bacteria have a nucleoid

    o Irregularly-shaped region that contains all or most of the genetic material

    o Not membrane-bound

        ▪ DNA not separated from the cytoplasm

- Key features of bacterial chromosomes

    o Chromosome is circular

        ▪ Usually, few million nucleotides in length

        ▪ Chromosome may be present in multiple copies

            • Depends on species and growth conditions

- - Have an origin of replication
    - Site where DNA replication is initiated
- Transcribed genes account for the majority of bacterial DNA
  - Non-transcribed DNA are "intergenic regions"
- For chromosomes to fit in a bacterial cell, DNA has to be compacted
  - Forms loop domains
    - Number of loops varies on the size of the chromosome and the species
  - DNA supercoiling
    - Supercoiling within loops compacts DNA further
- Supercoiling is accomplished by:
  - DNA gyrase (aka DNA topoisomerase II)
    - Creates negative supercoils
    - Can relax positive supercoils
  - DNA topoisomerase I
    - Relax negative supercoils

Eukaryotic Genomes

- Eukaryotic species contain one or more sets of chromosomes
  - Each set is composed of several different linear chromosomes
- Chromosomes located in the nucleus
  - Have to be highly compacted
    - Accomplished by binding of many protein
    - DNA-protein complex is called chromatin
- Eukaryotic genomes vary in size
  - Size difference is not really because of extra genomes
    - Really because of accumulation of repetitive DNA sequences that don't encode protein

Sequence Types (In Terms of Repetition)

- Unique or non-repetitive

  - Occur once or only a few times in the genome

  - Include transcribed and non-transcribed regions

- Moderately repetitive

  - Occur few hundred to a few thousand times

  - Include

    - Genes for rRNA and histones

    - Origin of replication

    - Transposable elements

- Highly repetitive

  - Occur several thousand or millions of times

  - Relatively short

  - Interspersed throughout the genome

Eukaryotic Chromatin Compaction

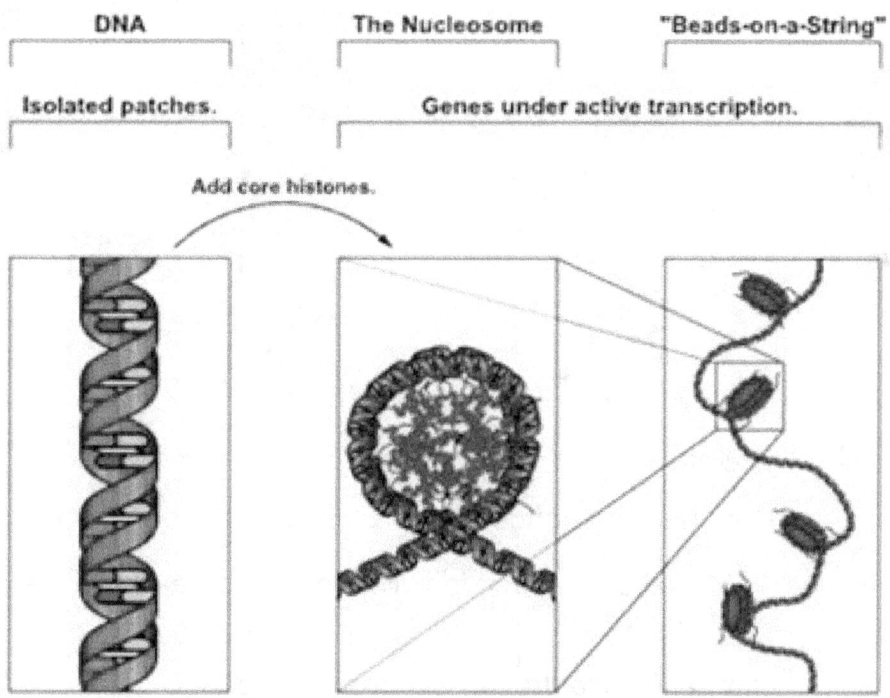

- Nucleosome - repeating structural unit within eukaryotic chromatin

- - - o Composed of double-stranded DNA wrapped around an octamer of histone proteins

    - Octamer consists of two copies of each core histone

- Five types of histones

    o Core histones: H2A, H2B, H3 and H4

        - Two of each make up the octamer

    o Linker histone: H1

        - Binds to linker DNA

        - Binds to nucleosomes

            - Not as tightly as are the core histones

- Nucleosomes aggregate to form the 30 nm fiber (a more compact structure)

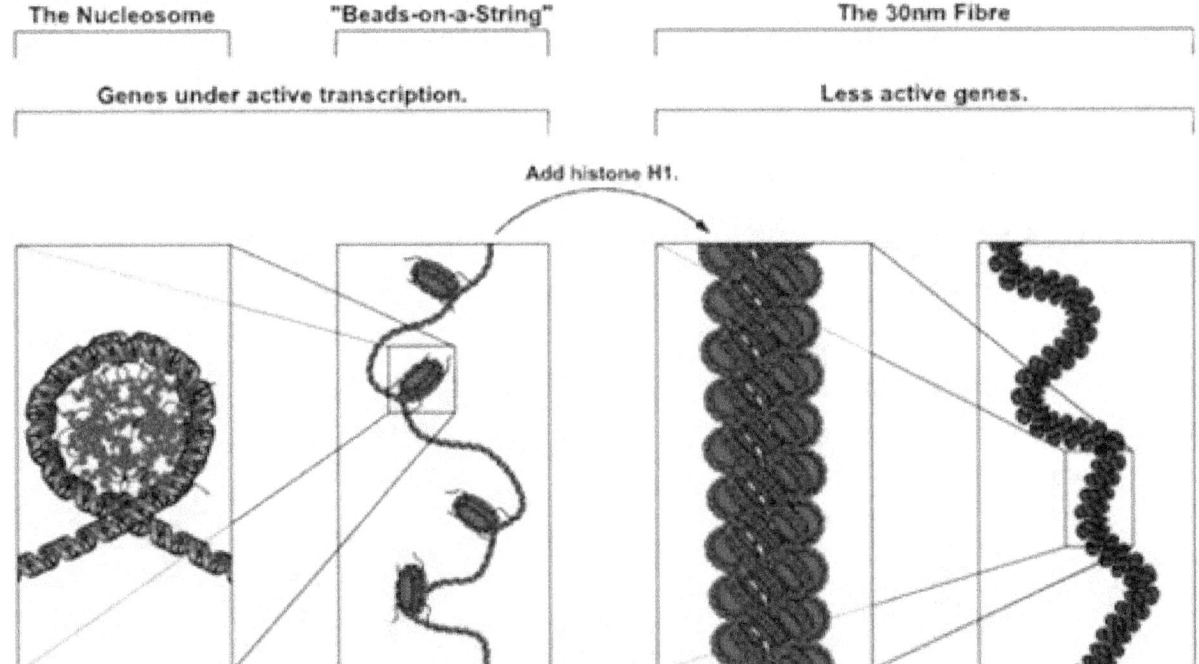

    o H1 is important in compaction

        - Moderate salt concentrations, H1 is removed

            - Less compact, look like "beads-on-a-string" (refer to the figure above)

        - Low salt concentrations, H1 remains bound

            - Creates greater compaction

- Third level of compaction occurs from the interaction of the 30 nm fiber with the nuclear matrix

    o Nuclear matrix has two parts

        ▪ Nuclear lamina

        ▪ Internal matrix proteins

            • 10 nm fiber and its associated proteins

- Radial loop domains form from the interaction between the 30 nm fiber and the nuclear matrix

    o Radial loops attaching to the nuclear matrix are involved in gene regulation

        ▪ Also help organize chromosomes within the nucleus

- Compaction level of chromosomes in interphase is not the same

    o Heterochromatin (HC)

        ▪ Densely packed; stains intensely

        ▪ Generally, no transcription

    o Euchromatin (EC)

        ▪ Less condensed; stains less well

        ▪ Transcribed at a higher rate

        ▪ Region where 30 nm fiber forms radial loop domains

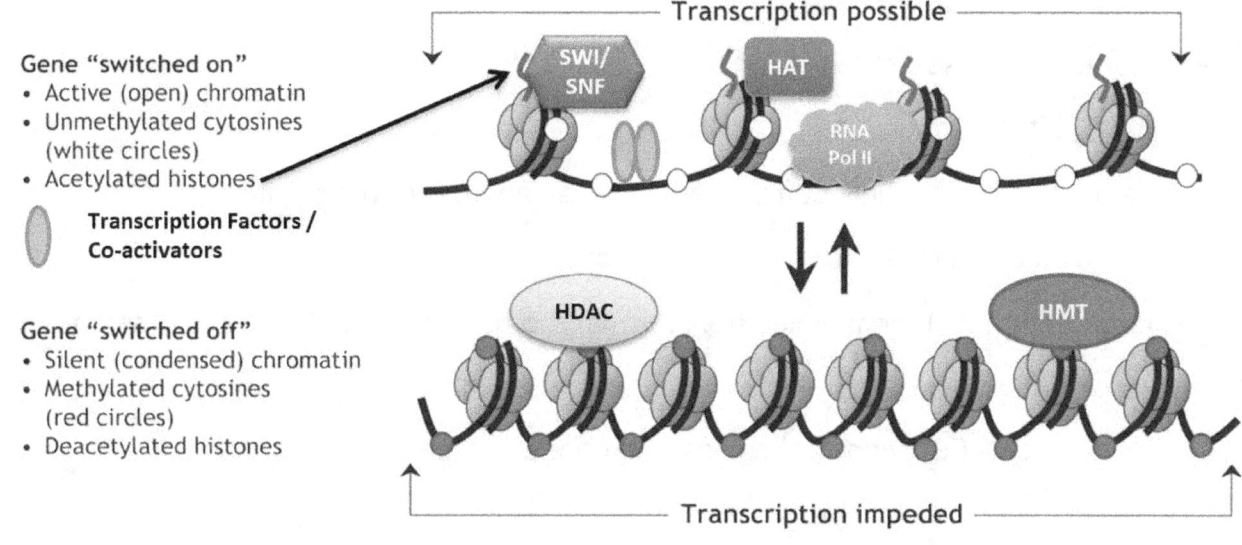

- Heterochromatin types

  - Constitutive

    - Always heterochromatic

    - No transcription; permanently inactive

  - Facultative

    - Interconverts between euchromatin and heterochromatin

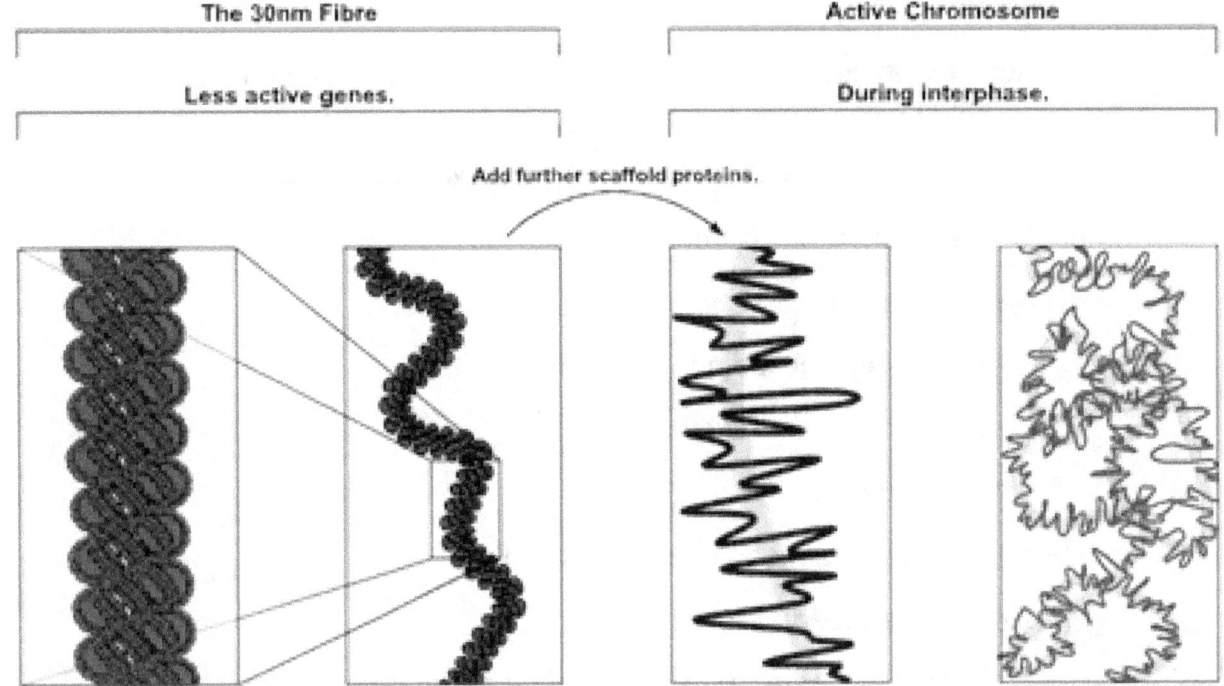

Metaphase Chromosomes

- Highly condensed chromosomes

  - Little transcription

- Radial loops are highly compacted and stay anchored to a scaffold

  - Scaffold forms from the nuclear matrix

- Two multi-protein complexes help form and organize metaphase chromosomes

  - Condensin

    - Function in chromosome condensation

- Cohesion
  - Function in chromatid alignment
- Both have SMC (structural maintenance of chromosome) protein
  - Catalyze changes in chromosome structure

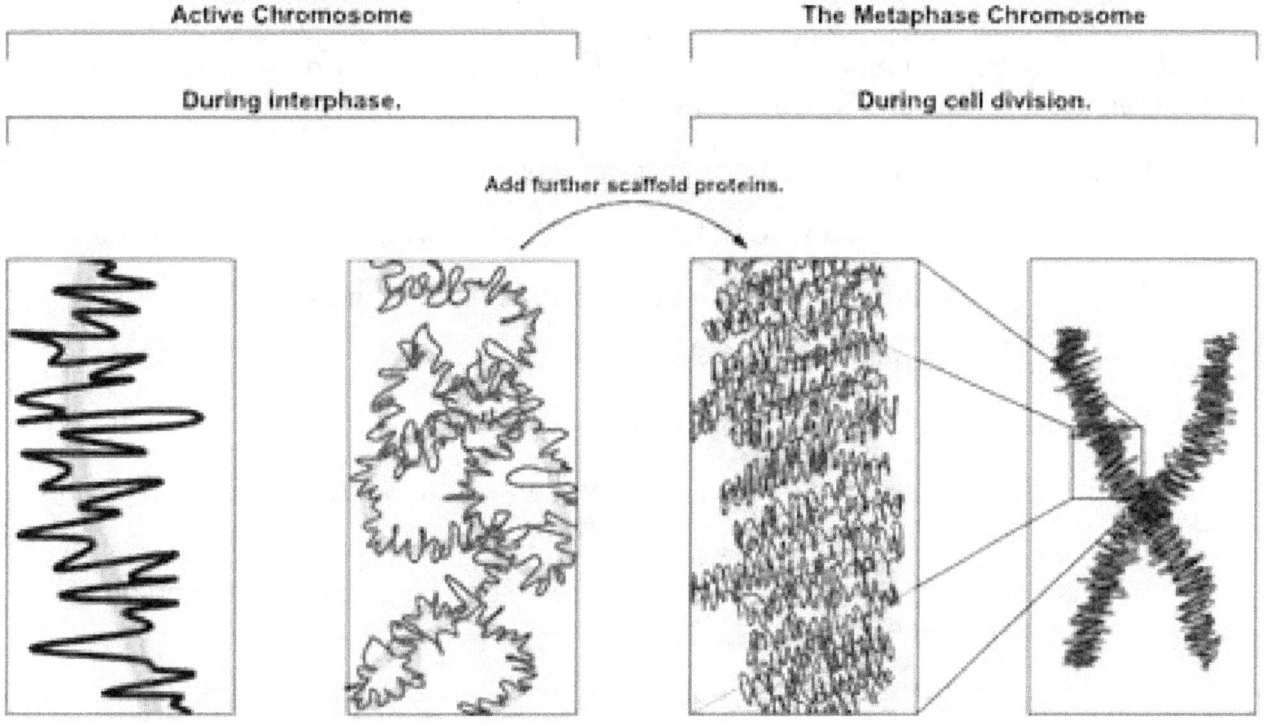

CHAPTER 12: DNA REPLICATION

DNA Replication Overview

DNA replication is the replication of a strand of DNA into two daughter strands, each of which consists of a strand from the original DNA double helix (semi-conservative mode of replication). A large number of enzymes and other proteins carry out DNA replication. DNA replication is different from transcription!

- In replication the two DNA strands come apart
    - Each serves as the template strand for the synthesis of new strands
        - Daughter strands – newly synthesized strands
        - Parental strands – original strands
- Semiconservative mode of replication
    - DNA is replicated in such a way that the double-stranded DNA contains one daughter and one parental strand
- In bacteria, the replication of DNA begins at a special site called the origin of replication (*Ori* site)
    - *Ori* site has a specific sequence of nucleotides recognized by replication enzymes
        - Enzymes separate the strands forming a replication "bubble"
    - Bacteria contain a single circular chromosome
        - Replication proceeds in both directions until the entire molecule is copied
- There may be numerous sites per chromosome in eukaryotic cells where replication can begin
    - At these origin sites, the DNA strands separate, forming a replication "bubble" with replication forks at each end

Replication in *E. coli*

E. coli are model organisms and their replication has been studied extensively.

- Origin of replication in *E.coli* is called *oriC*
- Significant DNA sequences in the chromosome of *E.coli*
    - AT-rich region
    - DnaA boxes
    - GATC methylation sites

Helicase, Topoisomerase, and Single-strand Binding Proteins

- Helicase unwinds the double helix and separates the two template DNA strands at the replication fork
    - Unwinding causes DNA ahead of the replication fork to become overwound
        - Topoisomerase helps relieve the strain from the DNA being overwound
- Single-strand binding proteins keep the unpaired template strands apart during replication
    - Also help protect the single stranded DNA inside the replication fork

DNA Polymerase and Primase

- DNA polymerases synthesize a new DNA strand at a replication fork
    - As nucleotides align with complementary bases along the template strand, they are added to the growing end of the new strand
- In *E. coli* there are 5 DNA polymerases
    - Two DNA polymerases are involved in replication (DNA pol I and DNA pol III)
        - DNA pol I replaces RNA primers with DNA
        - DNA pol III is composed of 10 subunits
            - α-subunit synthesizes DNA
            - Complex of all 10 is called the DNA pol III holoenzyme

- o Other three DNA polymerases are involved in repair and replication of damaged DNA
  - DNA pol II, DNA pol IV, and DNA pol V
- DNA polymerases cannot initiate synthesis of a new strand
  - o DNA polymerases can only add nucleotides to the 3' end of a growing DNA strand
    - The 3' end has a free hydroxyl group
    - For synthesis to begin in the first place, a primer has to be laid down (i.e. the primer is the initial nucleotide chain)
  - o Primer is a short stretch of RNA with an available 3' end that is synthesized by primase
    - Primase is an RNA polymerase
    - RNA is later degraded and replaced by DNA pol I
- After the primer is synthesized, DNA polymerase adds DNA nucleotides to the 3' end of the RNA primer and continues synthesizing the new strand
  - o Parent strands in DNA are antiparallel
  - o Along the parent strand that runs 3' to 5'
    - DNA polymerase has no trouble continuously synthesizing a new complementary strand
      - It can simply follow the new DNA being revealed by the unwinding action of helicase
    - DNA strand synthesized in this manner is called the leading strand
    - Leading strand only requires a single primer
  - o Other parental strand runs 5' → 3' and is called the lagging strand
    - DNA polymerase has to synthesize in the direction away from the new DNA being uncovered by the unwinding action of helicase
    - Replication cannot happen continuously because the DNA polymerase has to go back to start on the newly revealed DNA
    - This means that the synthesis of the lagging strand occurs as a series of short segments called Okazaki fragments
    - Also means that more than one RNA primer must be laid down for DNA polymerase to initiate synthesis from

- DNA pol I removes the RNA primers and fills in the resulting gap with DNA

  - Uses 3' to 5' exonuclease activity to digest the RNA

  - 5' to 3' polymerase activity to replace it with DNA

Processivity of DNA Polymerase III

Processivity refers to an enzyme's ability to catalyze consecutive reactions without releasing its substrate.

- Processivity of DNA pol III is due to several subunits that comprise the DNA pol III holoenzyme

  - β subunit is shaped like a ring

    - Works as a sort of clamp

  - γ subunit helps the β subunit initially clamp onto the DNA

  - Some of the other subunits are needed for the α and β subunits to function optimally

- If the β subunit is not present

  - DNA pol III falls off the DNA template after synthesizing only a few dozen nucleotides

  - Rate of synthesis is about 20 nucleotides per second

- If the β subunit is present

  - DNA pol III stays on the DNA template for ~50,000 nucleotides

  - Rate of synthesis is about 750 nucleotides per second

Replication Termination in *E. coli*

- Opposite *oriIC* are a pair of termination sequences (*ter* sequences)

  - Designated T1 and T2

- Tus protein binds to *ter* sequences

  - Stops movement of the replication forks

- Replication is terminated when the oppositely advancing forks meet at either T1 or T2

- Replication creates two intertwined molecules
    - Catenanes – intertwined circular molecules
    - Separated by topoisomerases

Proofreading Mechanisms

- DNA replication has a high degree of fidelity (exactness)
    - Mistakes are extremely rare (one in 10^8 nucleotides)
- Why is fidelity so high
    - Mismatched pairs are unstable
    - Configuration of DNA polymerase active site
        - DNA polymerase is unlikely to catalyze bond formation of mismatched pairs
    - DNA polymerase can excise any base that doesn't properly complement the parental DNA strand during synthesis and replace it with an appropriate base
        - Proofreading ability is the 3' to 5' exonuclease activity of DNA polymerase

Eukaryotic DNA Replication

Eukaryotic DNA replication is not as well understood as prokaryotic replication. This is due to eukaryotes having large linear chromosomes, tight packaging within nucleosomes, and the more complicated regulatory mechanisms.

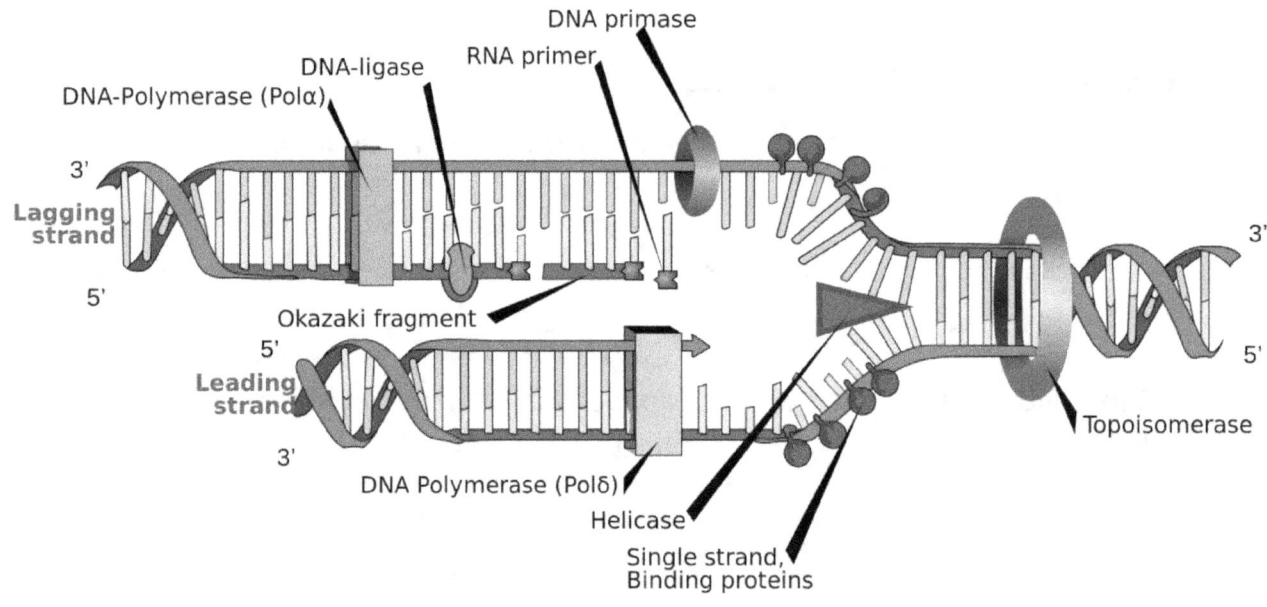

- Eukaryotes have multiple origins of replication
  - Due to having linear chromosomes instead of circular ones
- Origin replication complex (ORX)
  - Six-subunit complex
  - Initiates DNA replication

Polymerases and Replication

- In eukaryotes over 10 different DNA polymerases have been identified
  - Four of them function primarily in replication
  - Alpha (α), delta (δ), and epsilon (ϵ) are involved with nuclear DNA
  - Gamma (γ) is involved with mitochondrial DNA
- DNA pol α is the only one to associate with primase
  - DNA pol α and primase complex synthesizes a short RNA-DNA hybrid
    - Used by DNA pol δ and ϵ to elongate the leading and lagging strands

Polymerases and Repair

- DNA pol β is involved with base-excision repair
  - Removing incorrect bases from damaged DNA
- Lesion-replicating polymerases
  - Can synthesize a complementary strand over an abnormal region in the DNA

Nucleosomes and Replication

- Replication creates more DNA
  - More histones are needed
- Histones are synthesized in the S phase

Telomeres

A telomere is a region of repetitive nucleotide sequences at the end of chromosomes, which protects the ends of chromosomes from deterioration during DNA replication.

- Bacterial chromosomes are circular
    - Replication terminates where two replication forks meet
    - No need for telomeres
- Eukaryotic chromosomes are linear
    - No problem at 5' ends of the template strand
    - There is a problem at 3' ends of template strand
        - No way to replace the RNA primer with DNA
        - So each round of DNA replication leads to chromosome shortening
        - Creates a need for telomeres
- Telomerase
    - Adds sequence of 6 nucleotides to 3' end of DNA strand
        - Synthesis proceeds in 5' to 3' direction
        - Repeats of TTAGGG sequence (2-10 kb long)
    - Adds sequence of 6 nucleotides to 3' end of DNA strand
        - Uses RNA molecule (~451 nucleotides) as template
        - RNA is part of the structure of telomerase
        - Reverse transcriptase activity (RNA → DNA)
        - When 3' end is extended, DNA polymerase can then synthesize the complement
        - When replication ends there is still a shortened DNA segment, but the telomere is shortened and not the genome

- Telomere length may be a limiting factor in the life span of certain tissues and of the organism
    - May protect organisms from cancer
        - Limits the number of divisions that somatic cells can undergo
    - Active telomerase has been found in some cancerous somatic cells
        - Overcomes the shortening that would lead to self-destruction of the cancer

CHAPTER 13 – TRANSCRIPTION AND RNA MODIFICATION

Transcription Overview

Transcription is the process by which the information in a strand of DNA is copied into a new molecule of messenger RNA (mRNA). mRNA serves as a way of conveying genetic information from the DNA to the ribosome.

- RNA synthesis proceeds by complementary base pairing with one of the DNA strands

    o This strand acts as the template to determine which ribonucleotide gets added and in what order

    o This DNA strand guides the synthesis and is therefore the "noncoding" or template strand

- The complementary strand of DNA is therefore the coding strand

    o mRNA product is an RNA copy of the coding strand of the DNA

        ▪ With uracil in place of thymine

    o mRNA synthesized contains the genetic code

- Directions in transcription are referred to as "downstream" and "upstream"

    o Downstream – towards the 3' on the RNA

        ▪ Towards the 5' end on the template strand

    o Upstream – towards the 5' end on the RNA

        ▪ Towards the 3' end on the template strand

Genetic Code

The genetic code is a set of rules by which information encoded within genetic material is translated into proteins by living cells.

Seond letter

	U	C	A	G	
U	UUU ⎤ Phe UUC ⎦ UUA ⎤ Leu UUG ⎦	UCU ⎤ UCC ⎥ Ser UCA ⎥ UCG ⎦	UAU ⎤ Tyr UAC ⎦ UAA Stop UAG Stop	UGU ⎤ Cys UGC ⎦ UGA Stop UGG Trp	U C A G
C	CUU ⎤ CUC ⎥ Leu CUA ⎥ CUG ⎦	CCU ⎤ CCC ⎥ Pro CCA ⎥ CCG ⎦	CAU ⎤ His CAC ⎦ CAA ⎤ Gin CAG ⎦	CGU ⎤ CGC ⎥ Arg CGA ⎥ CGG ⎦	U C A G
A	AUU ⎤ AUC ⎥ Ile AUA ⎦ AUG Met	ACU ⎤ ACC ⎥ Thr ACA ⎥ ACG ⎦	AAU ⎤ Asn AAC ⎦ AAA ⎤ Lys AAG ⎦	AGU ⎤ Ser AGC ⎦ AGA ⎤ Arg AGG ⎦	U C A G
G	GUU ⎤ GUC ⎥ Val GUA ⎥ GUG ⎦	GCU ⎤ GCC ⎥ Ala GCA ⎥ GCG ⎦	GAU ⎤ Asp GAC ⎦ GAA ⎤ Glu GAG ⎦	GGU ⎤ GGC ⎥ Gly GGA ⎥ GGG ⎦	U C A G

First letter (rows) / Third letter (columns)

The Genetic Code:

- Triplet codons – nucleotide triplets that specify an amino acid

    o 64 codons

    o 20 common amino acids

    o 3 stop codons

- Code is degenerate

    o More than one codon for most amino acids

    o Third base "wobble"

    o Except methionine; always the start codon

- Code is universal

Bacterial Transcriptional Unit

Transcriptional unit refers to the part of the DNA that is actually transcribed into an RNA molecule.

- Important regions in bacterial DNA

    - Promoter

        - RNA polymerase binding site

        - Typically, located just upstream of the site where transcription begins

    - Terminator

        - Sequence that signals the end of a gene or operon

    - Regulatory sequences

        - Binding site for regulatory proteins

        - Regulatory proteins influence gene expression

- Important regions in bacterial mRNA

    - Ribosomal binding site (self-explanatory)

        - Translation begins close to this site

    - Start codon

        - Specifies the first amino acid in a protein sequence

        - Almost always a formylmethionine in bacteria

- Bacterial mRNA can be polycistronic

    - Polycistronic mRNA – mRNA which encodes for more than one polypeptide

Transcription in Prokaryotes vs. Eukaryotes

- In prokaryotes transcription occurs in the cytoplasm

    - Translation of mRNA to proteins also occurs in the cytoplasm

- In eukaryotes transcription occurs in the nucleus

    - mRNA then moves to the cytoplasm for translation

- DNA in prokaryotes is readily accessible to RNA polymerase

- DNA in eukaryotes is wrapped around proteins called histones

    o They order DNA into structural units called nucleosomes

    o Histones play a role in gene regulation and may not allow free access to a gene

- mRNA is not modified in prokaryotic cells

- Eukaryotic cells modify mRNA by RNA splicing, 5' end capping, and addition of a polyA tail

Transcription Steps

- RNA polymerase and cofactors unwind DNA strands at a specific point

    o RNA polymerases can only synthesis a new strand in the 5' → 3' direction

        ▪ Same as DNA polymerase

    o However, RNA polymerases don't need a free 3' end to start synthesizing the new strand

        ▪ Don't need a primer to start

    o Bacteria use only one RNA polymerase to synthesize all RNA molecules

    o Eukaryotes have three RNA polymerases (I, II, and III)

        ▪ RNA polymerase II participates in mRNA synthesis

- Specific sequences of nucleotides mark transcription start and end points

    o RNA polymerase attaches and initiates transcription at the promoter

        ▪ Promoter – region of DNA that initiates transcription

        ▪ Promoters are located near transcription start sites

    o Sequence signaling the end of transcription is called the terminator (in prokaryotes)

    o In prokaryotes RNA polymerase recognizes and binds directly to the promoter region

    o In eukaryotes proteins called transcription factors act as repressors or promoters in the recruitment of RNA polymerase

        ▪ Complex of transcription factors and RNA polymerase II bound to a promoter is called a transcription initiation complex

- An important promoter is the TATA box

  - Found in both eukaryotes and prokaryotes

  - Named so because contains 5'-TATAAA-3' sequence or a variant

- RNA polymerase unwinds the double helix 10 to 20 bases at time

  - Also adds nucleotides to the 3' end of the growing strand

  - One gene can be transcribed at the same time by several RNA polymerases

    - Allows a cell to make the encoded protein in large amounts

  - Transcription proceeds until RNA polymerase transcribes a terminator sequence in the DNA

    - RNA and DNA are released

- In eukaryotes, the pre-mRNA is cleaved from the growing RNA chain while RNA polymerase II continues to transcribe the DNA

  - The polymerase transcribes a DNA sequence called the polyadenylation signal sequence that codes for a polyadenylation sequence (AAUAAA) in the pre-mRNA

- At a point about 10 to 35 nucleotides past this sequence, pre-mRNA is cut

  - Polymerase continues transcribing for hundreds of nucleotides

  - Transcription is terminated once the polymerase eventually falls off the DNA

Promoters – In Detail

Promoters are regions of DNA that initiate transcription of genes.

RNA Polymerase, Sigma Factor (*E. coli Example*)

- Core enzyme

  - Contains four subunits: α, α, β, β'

- RNA polymerase holoenzyme

  - Core enzyme + sigma (σ) factor

- Sigma (σ) Factor
  - Smaller protein
  - Guides RNA Polymerase to target DNA sequence
  - Binds to promoters
    - -10 region (TATA box) & -35 region

Eukaryotic Promoters

- More complex promoters than prokaryotes
- Promoter elements
  - TATA: similar to prokaryotic TATA box
  - BRE: TFIIB recognition element
  - Inr: initiator sequence; contains transcription start site
  - MTE: motif ten element
  - DPE: downstream promoter element
  - Not all promoters contain all of these
  - Work synergistically
- Core promoter
  - Relatively short
  - Consists of TATA box
  - By itself is responsible for producing low levels of transcription
    - These low levels of transcription are called basal transcription
- Transcription Factors
  - General transcription factors (6)
  - TFIIA, TFIIB, TFIID, TFIIE, TFIIF, TFIIH
    - II: specific for RNA polymerase II
  - Bind to different elements in promoters
    - TFIIB binds to BRE; TBP (of TFIID) binds to TATA (TATA Binding Protein)

Eukaryotic Transcription

RNA Polymerases

- Nuclear DNA is transcribed by:
    - RNA polymerase I ⇒ makes rRNA
    - RNA polymerase II ⇒ makes mRNA
    - RNA polymerase III ⇒ makes tRNA and others

Regulatory Elements

- Affect binding of RNA polymerase to the promoter
- Two types
    - Enhancers
        - Stimulate transcription
    - Silencers
        - Inhibit transcription
- Typically, found in the -50 to -100 region
- *Cis*-acting elements
    - Non-coding DNA sequences which regulate transcription of nearby genes
- *Trans*-acting elements
    - Regulatory proteins that bind to *cis*-acting elements

Chromatin and Transcription

- Most transcription occurs in interphase
    - Chromatin is organized into the 30 nm fiber with radial loop domains during this phase
        - DNA is wound around the histone octamers and forms nucleosomes

- Histone octamer is about five times smaller than the RNA pol II and GTFs complex

    o Tight wrapping of DNA within the nucleosome inhibits the function of RNA pol

    o Chromatin structure is loosened during transcription

        ▪ Allows RNA polymerase to carry out its function

- Common mechanisms of altering chromatin structure

    o Covalent modification of histones

        ▪ Amino terminals of histones are modified by acetylation, phosphorylation, and methylation

        ▪ Histone acetyltransferase catalyzes the addition of acetyl groups

            • Adding acetyl groups loosens interactions between histones and DNA

        ▪ Histone deacetylase removes acetyl groups

            • Restores the tighter interaction between histones and DNA

    o ATP-dependent chromatin remodeling

        ▪ Energy from ATP is used to alter the structure of nucleosomes to make DNA more accessible

Post-transcriptional Modification

Eukaryotic cells modify RNA after transcription. Usually both ends of the RNA are altered.

- Transcription takes place in the nucleus

    o mRNA is processed before transport to cytosol for translation

- Processing mRNA is closely linked with transcription

- Modifications have important functions

    o Facilitate export of mRNA from the nucleus

    o Help protect mRNA from hydrolytic enzymes that cleave and degrade mRNA

    o Help ribosomes attach to the 5' end

Capping

- 5' cap

    o Modified form of guanine (7-methyl guanosine) is covalently added to the 5' end

    o Protects mRNA from exonuclease digestion

Tailing

- 3' tail

    o Sequences in the transcript (AAUAAA) signal for cleavage

    o Poly(A) polymerase generates a polyadenylate tail (~200 A residues)

    o Poly(A) binding protein binds to help protect this end from nuclease digestion

    o Half-life of mRNA depends on how rapidly the poly(A) tail is shortened

        ▪ Deadenylating exonucleases remove protective poly(A) tail

        ▪ Other enzymes clip off the 5' cap

        ▪ mRNA is then digested from both ends

Splicing

- Eukaryotic genes contain both exons (expressed regions) and introns (intervening regions that are not expressed)

- Splicing reactions cut out introns and connect exons

    o Starts even before transcription is finished

    o Creates an mRNA molecule with a continuous coding sequence

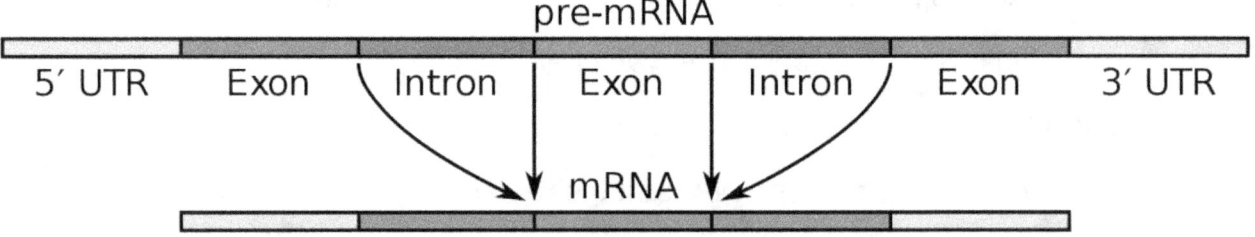

- Splicing mechanisms

    o All involve removing introns

    o All involve linking exons via a phosphodiester bond

- o Three mechanism types:
  - ▪ Group I intron splicing
  - ▪ Group II intron splicing
  - ▪ Spliceosome
- Splicing among group I and II introns is called self-splicing
  - o Occurs without enzymes
  - o RNA functions as its own ribozyme
    - ▪ Ribozyme – RNA molecule capable of acting as an enzyme
- Spliceosome
  - o Complex of 5 small RNA molecules and 100's of proteins that splice the mRNA
    - ▪ These are small nuclear RNAs (snRNA)
    - ▪ Catalyze chemical reactions that remove introns and covalently link exons
  - o Recognizes specific sequences at intron/exon junctions & branch point A
- Benefits of Splicing
  - o You can splice a DNA sequence in different ways
    - ▪ One gene allows for the production of multiple products

CHAPTER 14: TRANSLATION

Overview of Translation

Translation is the process of translating the language of RNA into the language of proteins.

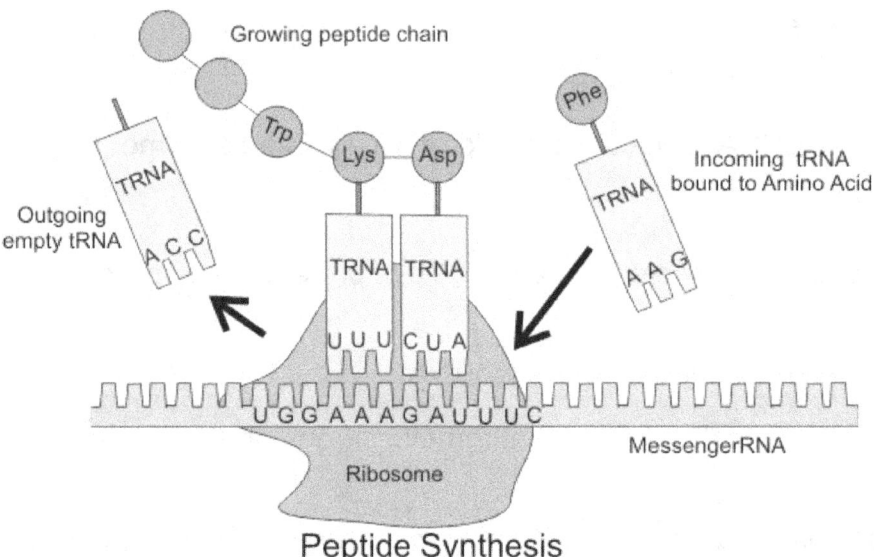

Peptide Synthesis

- Components

    - mRNA transcript - genetic code that is being translated

    - Ribosome - makes the protein

    - tRNA - amino acid carrier

    - Amino acids - what constitutes a protein

- 3 types of RNA participate in translation

    - mRNA (messenger RNA): contains the genetic code to specify the amino acid sequence of the protein

    - tRNA (transfer RNA): carries amino acids and reads the code

    - rRNA (ribosomal RNA): part of the ribosome that makes the protein

tRNAs

- tRNA molecules are transcribed from DNA templates in the nucleus

  - Have specific 2-D and 3-D shapes

  - Have a loop containing an anticodon and amino acid attachment site at the 3' end

- tRNAs have 3-base anticodon

  - Base pairs to codons in mRNA

    - Codon in 5' to 3' direction

    - Anticodon in 3' to 5' direction

  - Anticodons of some tRNAs recognize more than one codon

- tRNAs deposit amino acids in the prescribed order specified by the mRNA

  - Ribosome joins them into a polypeptide chain

  - tRNAs are used repeatedly

    - After depositing a specific amino acid at the ribosome from the cytosol, they return back to the cytosol to pick up that amino acid

- Amino acids are joined to the correct tRNA by aminoacyl-tRNA synthetase

  - 20 different synthetases for the 20 different amino acids

  - Synthetase has an active site that recognizes only a specific tRNA-amino-acid combination

    - Forms a covalent bond between them resulting in an aminoacyl-tRNA (activated amino acid)

Ribosome Structure

The ribosome has two major components, the small ribosomal subunit which actually reads the RNA and the large subunit which joins the amino acids together to form a polypeptide chain. Each of the subunits is composed of rRNA and a variety of proteins.

- 2 subunits, 52 proteins, 3 rRNAs

- Can bind 1 mRNA + 3 tRNAs

- Binding sites for tRNA

  o A site: carries tRNA with the next amino acid to be added

    ▪ Acceptor for the growing protein during peptide bond formation

  o P site: holds the growing peptide chain

  o E site (exit): discharged tRNAs leave the ribosome here

- mRNA makes a sharp bend between codons in A and P sites

  o Allows two tRNA molecules to fit side-by-side

  o May also prevent the ribosome from slipping

Translation Process

Translation can be divided into three stages: initiation, elongation, and termination.

Initiation

- mRNA, tRNA with the first amino acid, and the two ribosomal subunits come together

  o Small subunit binds with mRNA

- Small subunit moves downstream along the mRNA looking for the start codon (AUG)

- The initiator tRNA carrying methionine is already associated with the complex when the small subunit finds the start codon

  o Initiator tRNA hydrogen-bonds with the start codon

- Initiation factors bring the large subunit

  o Initiator tRNA occupies the P site

Elongation

- Elongation is a 3 step process

  o Codon recognition

  o Peptide bond formation

  o Translocation

- Codon recognition

  - Elongation factor assists hydrogen bonding between the mRNA codon under the A site and the corresponding anticodon

- Peptide bond formation

  - rRNA molecule catalyzes the formation of a peptide bond between the polypeptide in the P site and the new amino acid in the A site

    - Separates the tRNA at the P site from the growing polypeptide chain

    - Also transfers the growing chain to the tRNA at the A site

- Translocation

  - Ribosome moves the tRNA in the A site to the P site

  - The anticodon remains bonded to the mRNA codon causing the mRNA to move along with it

  - Next codon becomes available at the A site

  - tRNA that was in the P site is moved to the E site and leaves the ribosome

- The three steps of elongation continue to occur until completion of the polypeptide chain

Termination

- Occurs when one of the three stop codons reaches the A site

- Release factor binds to the stop codon and breaks the bond between the polypeptide and the tRNA in the P site

  - Frees the polypeptide

  - Translation complex disassembles

Comparing Gene Expression in Prokaryotes and Eukaryotes

Transcription and translation occur in very similar ways in both prokaryotes and eukaryotes, but the cellular machinery used and details vary between them.

- Eukaryotic RNA polymerases require transcription factors

  o Prokaryotic ones don't

- Transcription termination differ

- Ribosomes are different

- Prokaryotes can transcribe and translate the same gene at the same time

  o Eukaryotes cannot

    ▪ Nuclear envelope separates transcription from translation

CHAPTER 15: BACTERIAL GENE REGULATION

Overview of Gene Regulation

Gene regulation is important for many cellular processes (e.g., metabolism, cell division).

- Gene regulation – process of turning genes on and off

    o Ensure appropriate genes are expressed at the right time

- Constitutive gene – gene that is continuously transcribed (i.e. an unregulated gene)

    o Typically, gene product is needed at constant levels and is essential for survival

- Facultative gene – gene that is only transcribed when needed

Transcriptional Regulation

- Involves regulatory proteins

    o Repressors – inhibit transcription when they bind to DNA

    o Activators – promote transcription whey they bind to DNA

- Negative control – regulation by factors that block or turn off transcription

- Positive control – regulation by factors that are required for the activation of a transcriptional unit

- Transcription is also regulated by small effecter molecules

    o Don't bind to DNA

        ▪ Bind to the regulatory proteins

- Small effector molecule might increase transcription

    o Called "inducers" in this case

    o Function by either:

        ▪ Binding to an activator, making the activator bind to DNA

        ▪ Binding to a repressor, preventing it from binding to DNA

    o Genes regulated this way are said to be "inducible"

- Small effector molecule might decrease transcription
    - Called either "corepressors" or "inhibitors" depending on their mode of action
        - Corepressors bind to repressors, making the repressors bind to DNA
        - Inhibitors bind to activators, preventing them from binding to DNA
    - Genes regulated this way are said to be "repressible"

The *lac* Operon

The lac operon is required for the transport and metabolism of lactose in E.coli. An operon is a functioning unit of DNA that contains multiple genes transcribed from one promoter (operator region). All genes are transcribed together.

- Operon encodes for polycistronic mRNA
    - Contains the coding sequence for two or more genes
- Operon consists of a promoter, terminator, genes, operator
- DNA elements of the *lac* operon
    - Promoter region
        - Binds RNA polymerase
    - Operator region
        - Binds *lac* repressor protein
    - CAP site
        - Binds CAP (catabolite activator protein)
- Genes encoded by the *lac* operon
    - *lacZ* gene encodes β-galactosidase
        - Hydrolyzes lactose to galactose + glucose
        - Converts lactose into allolactose
    - *lacY* gene encodes lactose permease
        - Transporter for lactose and its analogues
    - *lacA* gene encodes thiogalactoside transacetylase
        - Function is unclear

- *lacI* gene encodes the *lac* repressor

  - Not part of *lac* operon

  - Blocks transcription

  - Repressor binds to operator

    - Blocks σ factor from binding promoter

  - Always present

    - Default expression is OFF for the *lac* operon

- *lac* operon can be regulated by:

  - Repressor protein

    - Inducible, negative control mechanism

    - Allolactose is the inducer

      - Binds to the repressor and inactivates it

  - Activator protein

    - Inducible, positive control mechanism

    - Small molecule involved is cAMP (cyclic AMP)

    - cAMP binds to CAP forming the cAMP-CAP complex

      - Complex binds to the CAP site and increases transcription

    - When glucose levels are high in the cell, cAMP levels are low

      - Means cAMP is not available to bind CAP

        - Transcription rate is decreased

- Mutants

  - I^- mutant – has a repressor that can't bind to the operator

    - *lac* operon is constitutively expressed

  - I^s mutant – has a repressor that can't be inactivated

    - No expression

  - O^C mutant – has an operator that the repressor can't bind to

    - *lac* operon is constitutively expressed

- ○ *lacP* mutant – RNA polymerase can't bind to the DNA
  - No expression

The *lac* Operon and its Control Elements

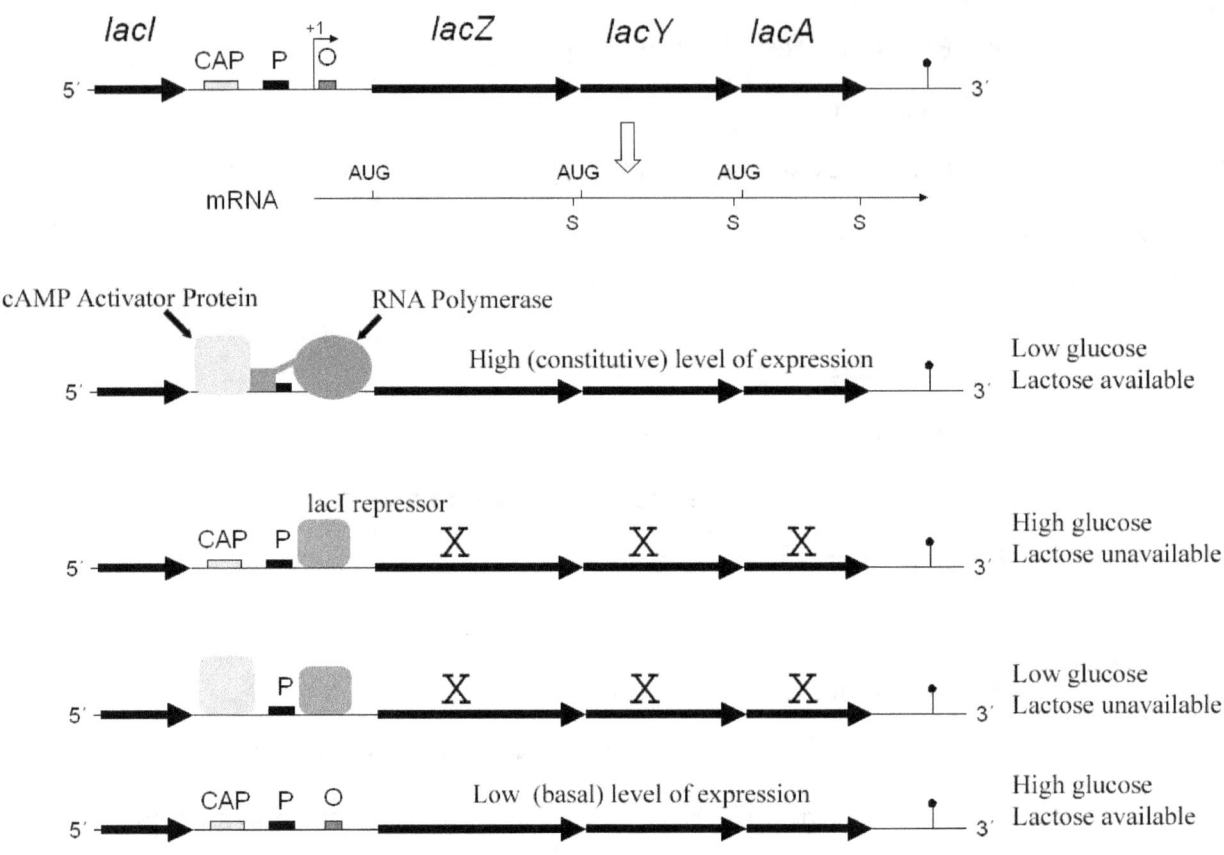

The *trp* Operon

The trp operon codes for the components involved in the production of tryptophan.

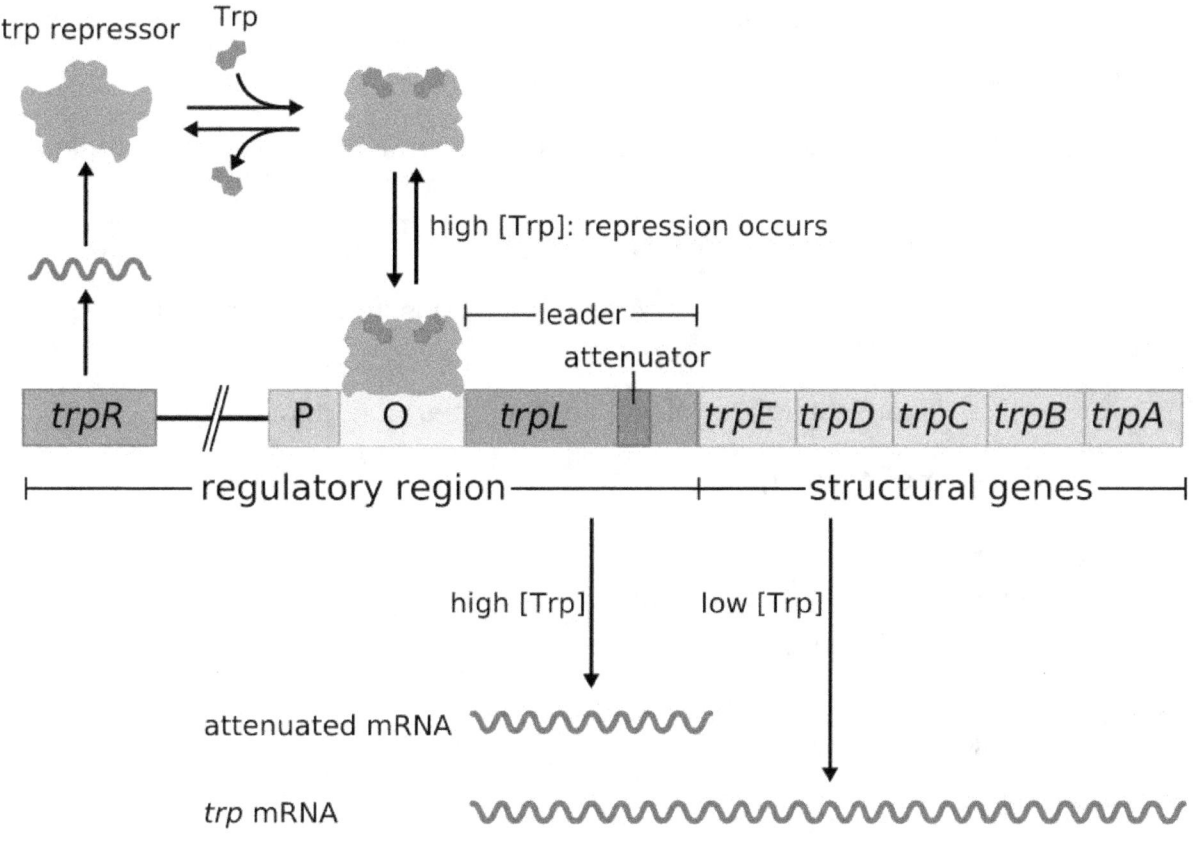

- *trpE, trpD, trpC, trpB* and *trpA* encode enzymes involved in the production of tryptophan

- *trpR* and *trpL* are responsible for regulation

   ○ *trpR* - encodes the *trp* repressor protein

   ○ trpL - encodes a short peptide called the Leader peptide

      ▪ Functions in attenuation

      ▪ Gene contains several trp codons (significance is explained farther on)

- At low tryptophan levels, the operon is transcribed

   ○ Repressor is inactive and doesn't bind to the operator site

- At high tryptophan levels, the repressor is activated

  ○ Tryptophan functions as a corepressor

    ▪ Don't need to produce tryptophan if tryptophan levels are already high in the cell

- At high tryptophan levels, attenuation occurs

  ○ Attenuations occurs because transcription and translation are coupled in bacteria

  ○ Transcription begins but it is terminated before the mRNA is made

    ▪ The attenuator (a segment of DNA) is important in facilitating termination

    ▪ Transcription terminates within the *trpL* region

High level of tryptophan

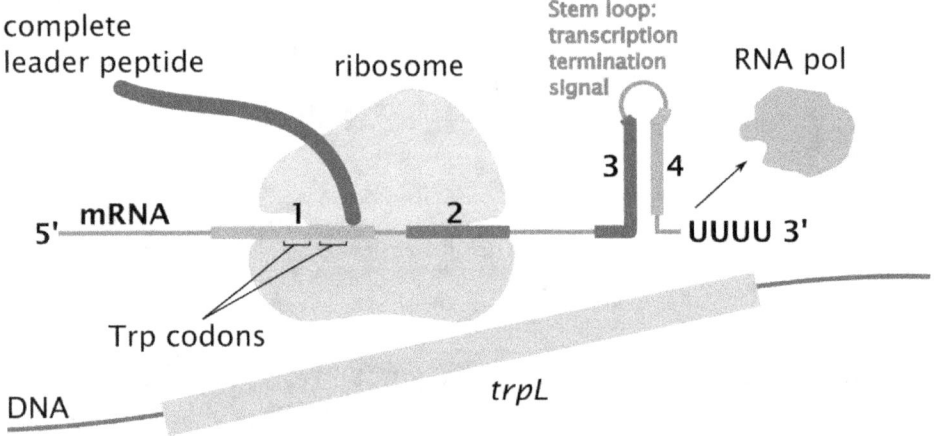

Low level of tryptophan

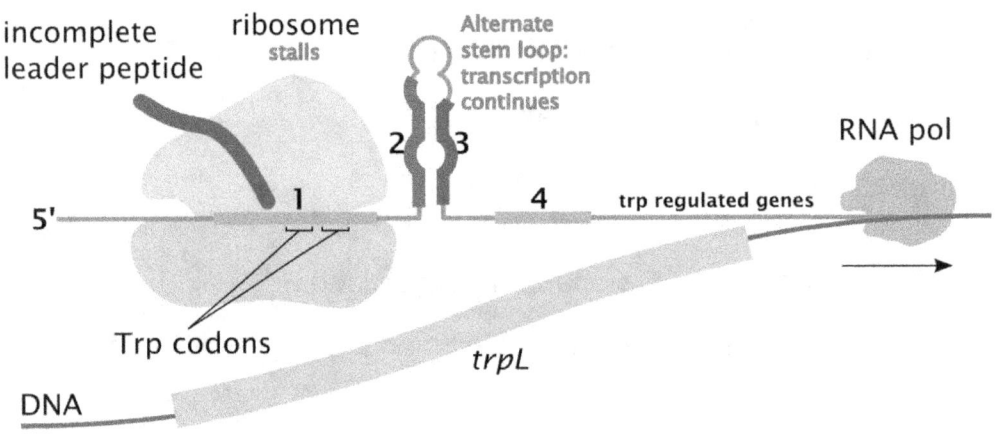

- o Region 2 is complementary to regions 1 and 3
- o Region 3 is complementary to regions 2 and 4
- o Means that many different stem-loops can form
- Formation of the 3-4 stem-loop causes RNA pol to terminate transcription at the end of the *trpL* gene

What happens at low levels of tryptophan?

- Ribosome stalls in the *trpL* region (region 1)
  - o Remember that there are several trp codons in this region
  - o Tryptophan levels are low in the cell
    - So there are low levels of charged trp-tRNAs as well
  - o When the ribosome starts to transcribe the *trpL* gene it stalls because it needs charged trp-tRNAs to proceed with translation
- Since region one is occupied, a 2-3 stem-loop forms
  - o Region 3 becomes occupied and the 3-4 stem loop can't form
  - o RNA pol transcribes the rest of the operon

What happens at high levels of tryptophan?

- Ribosome does not stall in the *trpL* region
  - o Plenty of charged trp-tRNAs because of the high levels of tryptophan in the cell
- Ribosome rapidly covers both region 1 and region 2
  - o Region 2 can't base pair with region 1 or region 3
- 3-4 stem-loop forms
  - o It is a transcription termination sequence
  - o RNA polymerase disassociates
  - o Transcription is attenuated

Translational and Posttranslational Regulation

Translational Regulation

- Translational regulatory protein recognize sequences within the mRNA
    - Typically, inhibit translation and are called "translational repressors"
    - Translational repressors inhibit translation by:
        - Binding next to the Shine-Dalgarno sequence (a ribosomal binding site in prokaryotic mRNA) or the start codon
            - Hinders ribosome from initiating translation
        - Binding outside the start codon region
            - Stabilizes mRNA secondary structure to prevent initiation of translation
- Another way to regulate translation is through the synthesis of antisense RNA
    - Antisense RNA is an RNA strand complementary to mRNA
    - Antisense RNA base pairs with the mRNA and physically obstructs it from being translated

Posttranslational Regulation

- Common mechanism is feedback inhibition
    - Final product of a pathway inhibits an enzyme that functions earlier in the pathway
    - Enzyme inhibited is an allosteric enzyme with two binding sites
        - Catalytic site – binds substrate
        - Regulatory site – binds final product of the pathway
    - If the concentration of the final product is high it will bind to the enzyme and inhibit the enzyme's ability to bind the substrate
- Another mechanism is covalent modification
    - Can be irreversible
        - Example: attachment of prosthethic groups, sugars, lipids
    - Can be reversible
        - Example: phosphorylation, acetylation, methylation

CHAPTER 16: EUKARYOTIC GENE REGULATION

Introduction to Gene Regulation in Eukaryotes

- Benefits

    o Allows for eukaryotes to respond to changes in nutrient availability

    o Allows them to respond to environmental stimuli

- Necessary for

    o Proper expression of genes during the various stages of the life cycle

    o Different expression of genes in the distinct cell types

Regulatory Transcription Factors

Transcription factors are proteins that are involved in the process of transcribing DNA into RNA.

- Two main types

    o General transcription factors (GTFs)

        ▪ Needed for RNA pol to bind to the core promoter

        ▪ Needed for basal level transcription

    o Regulatory transcription factors (rTFs)

        ▪ Regulate the rate of transcription of nearby genes

        ▪ Influence the ability of RNA pol to begin transcription

- rTFs recognize response elements (aka control elements) near the core promoter

    o Binding of rTFs to them affects transcription of the associated gene

    o Typically, located a few hundred nucleotides upstream of the promoter

        ▪ However, can be at other sites

            • Downstream of the promoter

            • Within introns

            • Thousands of nucleotides away

- Transcription factor have domains that facilitate specific functions

  - Example: domains can be for DNA-binding or be binding site for effector molecules

- Motif - domain that has a very similar structure in many proteins

Enhancers and Silencers

- Activator – regulatory protein that increases the rate of transcription

  - Enhancer - sequence the activator binds to

- Repressor - regulatory protein that decreases the rate of transcription

  - Silencer - sequence the repressor binds to

- Binding of an activator to an enhancer increases the rate of transcription

  - Known as up-regulation of a gene

- Binding of a repressor to a silencer decreases the rate of transcription

  - Known as down-regulation of a gene

TFIID and Mediator

- Usually, regulatory transcription factors don't bind directly to RNA polymerase

  - Two common protein complexes that communicate the effects of the regulatory transcription factor are TFIID and mediator

- TFIID

  - GTFs bind to TATA box

  - Recruit RNA polymerase to the core promoter

  - Activator recruits TFIID

    - Transcription activated

  - If a repressor is present instead of an activator, the repressor inhibits TFIID binding

    - Transcription repressed

- Mediator
    - Activator stimulates the function of the mediator
        - Allows RNA pol to form preinitiation complex
        - RNA pol proceeds with transcription
    - Repressor inhibits the function of the mediator
        - Transcription is repressed

Steroid Hormones

- Steroid receptors are regulatory transcription factors that respond to steroid hormones
- Steroid hormones are produced to affect gene transcription
    - Steroid hormones are produced by endocrine glands
        - Secreted into the bloodstream
        - Then taken up by cells
- Glucocorticoids
    - Influence nutrient metabolism
        - Promote glucose utilization
        - Promote fat mobilization
        - Promote protein breakdown
- Gonadocorticoids
    - Examples: estrogen and testosterone
    - Influence growth and function of gonads (organs that produce gametes)

DNA Methylation

DNA methylation is the process by which methyl groups are added to DNA. This typically silences gene expression, especially when it occurs near the promoter. DNA methylation is heritable; inherited during cell division.

- Many genes in vertebrates and plants have CpG islands near their promoters
    - CpG is shorthand for cytosine and guanine separated by a single phosphate

- Housekeeping genes – constitutive genes required for maintenance of basic cellular function
    - CpG islands are unmethylated in these genes
- Tissue-specific genes
    - CpG islands are methylated to silence unneeded genes

Regulation of RNA Processing and Translation

RNA Processing Regulation

- Alternative splicing
    - Pre-mRNA can be spliced in various ways
        - Allows for the production of more than one protein
    - Cell specific
        - Protein can be modified to function more effectively in a particular type of cell
    - Involves proteins called splicing factors
        - Have controlling influence on spliceosome's ability to recognize or choose splice sites
        - Can inhibit or enhance the spliceosome's ability to recognize
- RNA editing
    - Changes in the nucleotide sequence of an RNA molecule
        - Can be addition or deletion of bases
        - Or conversion of one base to another base
- Stability of mRNA
    - Can be regulated to increase or decrease the half-life of the mRNA
        - Effects mRNA concentration
            - Which effects gene expression

- Stability can be affected by the length of the polyA tail
  - Overtime the tail is shortened by cellular nucleases
  - PolyA binding proteins bind to the tail to stabilize it
    - Can't bind when the tail gets too short
- RNA interference of double-stranded RNA
  - Can silence expression of genes
  - Mediate degradation of homologous mRNAs

Translation Regulation

- General regulation of translation
  - Regulating the function of translational initiation factor permits or inhibits translation
  - Useful to inhibit translation if a cell is exposed to a virus or a toxic substance
    - Virus won't be able to manufacture viral proteins
- Translational regulation of specific mRNAs
  - Some mRNA can be regulated by binding proteins that inhibit ribosomes from initiating translation
  - Usually, the proteins bind to 5' end of the mRNA to prevent the ribosome from binding

CHAPTER 17: MUTATIONS AND DNA REPAIR

Mutation

Mutations are changes in the structure of a gene, resulting in a variant form that may be transmitted to subsequent generations. Mutations are caused by the alteration of single base units in DNA, or the deletion, insertion, or rearrangement of larger sections of genes or chromosomes.

- Mutations that change the base sequence of DNA may be neutral (silent), beneficial (adaptive), or harmful (deleterious)

    o Neutral mutations neither benefit nor damage the organism

    o Adaptive mutations could lead to antibiotic resistance or altered pathogenicity

    o Deleterious mutations lead to loss of function, the formation of a cancer, or death

- Mutations can be harmful so organisms have developed ways to repair damaged DNA

Types of Mutations

- Silent mutations – mutations that do not alter the amino acid sequence of the protein

- Missense mutations – mutations that do cause a change in the amino acid sequence

- Nonsense mutations – mutations that result in an early stop codon

- Frameshift mutations – involve the deletion or addition of one or two amino acids and their multiples

    o Causes a shift in the reading frame

- Point mutation (substitution) – single nucleotide substitution

    o Two types

        ▪ Transition: purine changed to another purine or pyrimidine changed to another pyrimidine

        ▪ Transversion: purine changed to pyrimidine or vice versa

- Deletion – deletion of at least 1 nucleotide

- Insertion – insertion of at least 1 nucleotide
- Mistakes by DNA polymerase
  - Mis-matched bases
- Damage by ROS (Reactive Oxygen Species) like $\cdot O_2^-$ or H_2O_2
  - By-products of oxidative metabolism
  - Example: G oxidized to oxoG
    - Can base pair with either C or A
- Spontaneous depurination
  - Glycosidic bond connecting base to sugar is broken
  - Results in abasic site
  - Occurs ~18,000X/day
- Spontaneous deamination
  - Removal of an amine group
  - Original G:C base pair can become T:A
  - Occurs ~500 times per day
- Germ-line mutations occur directly in a sperm or egg cell
  - Or in a precursor cell
- Somatic mutations occur directly in a body cell
  - Or in one of their precursor cells

Effects on Genotype and Phenotype from Mutations

- Wild-type alleles – the most prevalent alleles in a population
  - Typically encode proteins that function normally and are produced in the proper amount
- Forward mutations – change a wild-type allele into a new allele (mutant allele)
- Reverse mutations – change a new allele back into the wild-type
  - Also called reversions

- Mutations that alter an organism's phenotype are called variants
    - Categorized by the effect they have on survival
        - Deleterious mutations – decrease chance of survival
        - Beneficial mutations – enhance the chance of survival
- Some mutations are conditional mutants
    - Affect phenotype only under specific conditions
- A second mutation can affect the phenotypic expression of another mutation
    - Second-site mutations are called suppressor mutations
    - Two types
        - Intragenic suppressors
            - Second-site of mutation is within the same gene as the first
        - Intergenic suppressors
            - Second-site of mutation is in a different gene from the first

Mutations in Noncoding Sequences

Mutations in noncoding sequences can still have an effect on gene expression.

- Mutation may cause a change in a promoter
    - Up promoter mutations increase the rate of transcription
    - Down promoter mutations decrease the rate of transcription
- Mutation may cause a change in response element or operator site
    - Might disrupt ability of a gene to be properly regulated
- Mutation may cause a change in 5'-UTR/3'-UTR
    - Might affect the ability of mRNA to be translated or its stability
- Mutation may cause a change in the splice recognition sequence
    - Might affect proper splicing of pre-mRNA

Causes of Mutations

Spontaneous Mutations

Spontaneous mutations are mutations that result from abnormalities in cellular processes. Example of spontaneous mutation is any error that occurs in DNA replication. Arise from different types of chemical changes:

- Depurination – removal of a purine from DNA

    o Covalent bond between deoxyribose and a purine is slightly unstable

    o Sometimes undergoes a spontaneous reaction with water

       ▪ Cause the base to be detached from the sugar

       ▪ These are apurinic sites

- Deamination

    o Deamination of cytosine (i.e. removal of an amino group from cytosine)

       ▪ Harder to deaminate the other bases

       ▪ Deamination of cytosine results in uracil

Deamination of Cytosine to Uracil: **Cytosine** → **Uracil**
(+ H_2O, – NH_3)

       ▪ DNA repair enzymes recognize uracil as a wrong base in DNA and can remove it

o Deamination of 5-methyl cytosine may also occur

▪ Results in thymine

Deamination of 5-methyl Cytosine to Thymine:

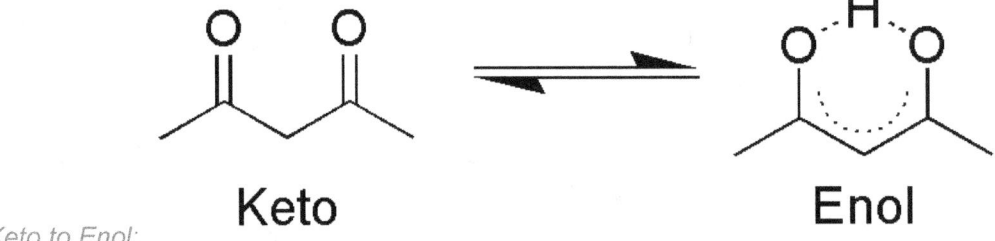

Cytosine   5-methyl   Thymine
             Cytosine

▪ Thymine is present in normal DNA

• Repair enzymes can't determine when an incorrect base occurs

• Tautomeric shift – spontaneous isomerization of a nitrogen base to an alternative hydrogen-bonding form

o Keto form is the common and stable form of both thymine (T) and guanine (G)

▪ T and G interconvert to an enol form at a low rate

Keto   Enol

Keto to Enol:

o Amino form is the common and stable form of both adenine (A) and cytosine (C)

▪ A and C interconvert to an imino form at a low rate

o Rare forms promote A to base pair with C and G to base pair with T

▪ Remember: normally A pairs with T and G pairs with C

o Mutations from tautomeric shift have to occur immediately prior to DNA replication

Induced Mutations and Mutagens

Induced mutations are mutations resulting from physical or chemical agents. Agents that cause alteration in DNA structure are called mutagens. Mutagens are classified as:

- Chemical - chemical substances that cause changes in the structure of DNA
    - Base modifiers – covalently modify structure of a nucleotide
        - Example: nitrous acid replaces amino groups with keto groups
    - Base analogues – structurally similar to DNA
        - Become accidently incorporated into daughter strands during DNA replication
    - Intercalating agents – flat planar structures that insert themselves into the double helix
        - Distorts the helical structure of DNA
            - Cause many problems during replication
- Physical - physical phenome that cause changes in the structure of DNA (e.g., X-rays, gamma rays, and UV light)
    - Ionizing radiation
        - Includes X rays and gamma rays; short wavelength and high energy
            - Can penetrate deeply
            - Create chemically reactive molecules (free radicals)
            - These can cause: base deletions, single nicks in DNA strands, cross-linking, and chromosomal breaks
    - Nonionizing radiation
        - Includes UV light; has less energy
            - Can't penetrate deeply
            - Causes formation of thymine dimers
                - Cause mutations when DNA is replicated

DNA Repair

There are many DNA repair systems within cells that are used to fix alteration within their DNA.

Direct Repair

- Photolyase – enzyme that repairs thymine dimers

    o Able to split the dimers

- O^6-alkylguanine alkyltransferase – enzyme that repairs alklylated bases

    o Transfers methyl or ethyl group from the base to its own cysteine side chain

        ▪ Permanently inactivates itself when it does this

Base Excision Repair (BER)

- Enzymes known as DNA N-glycoslyases are involved in this repair

    o Able to recognize abnormal bases and able to cleave the bond connecting an abnormal base from its sugar

- If a purine is removed, site is call apurinic

- If a pyrimidine is removed, site is called apyrimidinic

- Mechanism of repair

    o N-glycosylase recognizes an abnormal base

        ▪ Cleaves bond between it and its sugar

    o AP endonuclease recognizes the missing base

        ▪ Cleaves DNA backbone at the 5' side of the missing base

    o DNA polymerase uses its 5' to 3' exonuclease activity to remove the damaged region

        ▪ Synthesizes normal DNA in the region

    o Ligase connects the new region with the old

Nucleotide Excision Repair (NER)

- Can be used to repair thymine dimers, chemically modified bases, and missing bases

- Found in both prokaryotes and eukaryotes

- Example of NER in *E. coli*

    o Four important proteins involved: Uvr A, UvrB, UvrC, and UvrD

    ▪ Involved in the UV repair of pyrimidine dimers

    o Mechanism

    ▪ UvrA-UvrB complex searches for damaged DNA

    • Once the damage is found UvrA leaves and UvrC binds

    ▪ UvrC makes cuts on both sides of the thymine dimers

    ▪ UvrD, a helicase, removes the damaged DNA

    • UvrB and UvrC released

    ▪ DNA polymerase uses its 5' to 3' exonuclease activity to remove the damaged region

    • Synthesizes normal DNA in the region

    ▪ Ligase connects the new region with the old

Mismatch Repair

- DNA polymerases have 3' to 5' exonuclease activity

    o Proofreading ability to detect mismatched bases and fix them

- Proofreading can fail

    o Methyl-directed mismatch repair can be used to fix the mistake

- Example of mismatch repair in *E. coli*

    o MutL, MutH, and MutS are the primary proteins involved

- MutH is able to distinguish between a parental strand and the daughter strand immediately after replication

    - Before replication both strands are methylated

    - Immediately after replication

        - Parent strand is methylated

        - Daughter hasn't been methylated yet

- Mechanism

    - MutS finds mismatch

        - MutL binds to MutS

    - MutS-MutL complex binds to MutH

        - MutH is already bound to the hemimethylated sequence

    - MutH cuts the nonmethylated strand

    - MutU separates the DNA strands at the cleavage site

        - Exonuclease digests the nonmethylated strand past the site of the mismatched base

    - DNA polymerase synthesizes normal DNA in the region

    - Ligase connects the newly synthesized region with the old

CHAPTER 18: BIOTECHNOLOGY, RECOMBINANT DNA, GENOMICS

Introduction to Biotechnology

Biotechnology is the use and manipulation of microorganisms, cells, or cell components to make desired products (antibiotics, vitamins, etc.).

- Genetic engineering - process of using recombinant DNA technology to create new cells that produce chemicals that an organism doesn't naturally make

- Recombinant DNA technology - insertion or modification of genes in an organism to produce desired proteins/enzymes

    o Genes from one organism can placed inside another organism's DNA

        ▪ Including organisms of different species

    o Examples:

        ▪ Human gene for insulin production has been inserted into bacteria and they are able to produce human insulin for commercial use

        ▪ A gene that codes for a viral coat protein from the hepatitis B virus has been inserted into yeast for the commercial production of a vaccine against this disease

Overview of Recombinant DNA Procedures

- Gene of interest is inserted into a self-replicating vector (plasmid, transposon, viral DNA)

    o Accomplished through the use of restriction enzymes which can cleave DNA molecules at restriction sites

- Recombinant vector is then taken up by a cell

- The cell is then allowed to divide producing a culture

    o All cells in the culture will be identical

        ▪ Each carries the vector with the gene of interest

    o The gene of interest can be isolated in large quantities for further experimentation

- Expression of the gene of interest within the cells can produce a large volume of product (protein, enzyme, hormone, etc.) that can then be harvested

Tools of Biotechnology

- Selection - culturing of a naturally-occurring microbe that produces a desired product

- Site-directed mutagenesis - changing a specific gene to change a protein

Restriction Enzymes

- Cut specific sequences of DNA

- Naturally occur in some species of bacteria

  - Destroy bacteriophage (viral) DNA that gets into these species' cells

  - Cannot digest the bacterial (host) DNA because of the presence of methylated cytosines

- Purified forms of bacterial restriction enzymes are used in genetic engineering

- A specific restriction enzyme always recognizes and cuts DNA at a very specific nucleotide sequence of the DNA molecule

- Cuts made by some restriction enzymes are staggered, producing "sticky ends"

  - Cuts that are made, are the same on both strands of the DNA

  - However, they run in opposite directions

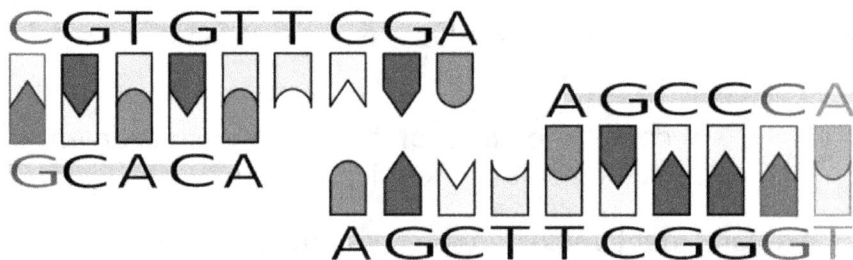

Example of "Sticky Ends":

  - Sticky ends can hydrogen bond with a complimentary base sequence

  - If two fragments of DNA from different sources have been cut with the same restriction enzyme, the two fragments will have complimentary sticky ends and can recombine

- DNA ligase can then be used to covalently link the sugar-phosphate backbones

  - Produces a recombinant DNA molecule

Vectors

- Transport DNA into the desired cell

- Plasmids, transposons, and viruses are potential vectors

- Characteristics of good vectors

  o Must be self-replicating

  o Should resist destruction by the recipient cell

    ▪ Circular vectors are highly resilient

    ▪ Viral DNA that inserts itself quickly into the host chromosome is more likely to remain intact

  o Should carry a marker gene

    ▪ Makes it easy to retrieve clones containing the vector

    ▪ For example, a marker gene may code for the production of a specific enzyme, or antibiotic resistance

- Shuttle vectors are plasmids capable of existing in several different species

  o Used to move genes from one species to another

- Vectors can be used to insert functional genes into human cells that have defective genes

  o This is gene therapy

Polymerase Chain Reaction (PCR)

- Used to make multiple copies of a piece of DNA

- Also used to:

  o Amplify DNA for recombination experiments

  o Sequence DNA

  o Diagnose genetic diseases

  o Detect pathogens

PCR Steps

- DNA of interest, RNA primers, DNA nucleotides, and DNA polymerase are placed in a thermocycler

- Contents are incubated at 94°C for a minute

    o Strands of the original piece of DNA separates

- Temperature is then lowered to 60°C for a minute

    o RNA primers anneal to the DNA

- Temperature is then increased to 72°C for a minute

    o DNA polymerase synthesizes a complimentary strand of DNA

- Cycle is repeated several times, resulting in an exponential increase in the number of DNA pieces

    o All of which are identical to the original piece of DNA

- DNA polymerase from the thermophilic bacterium *Thermus aquaticus* is used because of the high temperatures involved

    o This polymerase is able to function at high temperatures which would denature normal polymerases

- PCR can only be used to amplify small specific sequences of DNA as determined by the primer used.

    o Cannot be used to amplify an entire genome

Techniques of Genetic Engineering

Recombinant DNA in the lab is made outside a living cell. Once the recombinant DNA has been made, it must be put back into a cell. DNA can be inserted into a cell by:

Inserting DNA into a Cell

- Transformation

    o Some species spontaneously take up a recombinant plasmid and integrate it into their chromosomes by recombination

    o Many species are unable to spontaneously transform

        ▪ They first must be treated with chemicals to make them "competent" (i.e. able to take up external DNA)

- Electroporation
    - Application of an electrical current to a cell forms small pores in the plasma membrane that the DNA can pass through
        - Cells with walls must first be converted to protoplasts or spheroplasts (Gram+ and Gram- cells, respectively, without their cell walls) for this technique to work
- Protoplast fusion
    - Protoplasts in solution will fuse to form hybrid cells at a slow rate
        - Adding polyethylene glycol increases the rate of fusion
    - Once fusion occurs, the two chromosomes may undergo natural recombination
- Gene gun
    - Microscopic particles of gold or tungsten are coated with DNA and shot out of a gene gun with a burst of helium
- Microinjection
    - A glass micropipette is used to inject DNA into a cell

Obtaining DNA

- Two main sources of genes:
    - Gene libraries are made up of pieces of an entire genome stored in bacterial plasmids or in phages
    - Synthetic DNA is made by a DNA synthesis machine

Cloning the Genes of Eukaryotes

- Genes containing introns are not useful for genetic engineering
    - Makes them too large to work with
    - Introns will not be removed from the mRNA transcript in bacteria and the product will be non-functional
- Complementary DNA (cDNA) is made from an mRNA template by the enzyme reverse transcriptase
    - Result is a DNA molecule without introns
    - Commonly used to obtain eukaryotic genes that can then be inserted into a prokaryote

Selecting for Cells

Out of millions of cells, only a few may have successfully taken up the gene of interest.

Blue-White Screening

- Use a plasmid vector that contains a gene that codes for a chosen antibiotic resistance with the gene of interest
    - Host bacterium cannot survive on a medium that contains the antibiotic unless it has taken up the plasmid
- Plasmid vector has a second gene that codes for the enzyme Beta-galactosidase
    - Enzyme cleaves lactose into glucose and galactose
    - Beta-galactosidase gene has several sites that can be cut by restriction enzymes
        - Plasmid vector and foreign DNA containing the gene of interest are digested with the same restriction enzyme
    - Fragments of the foreign DNA may insert into the Beta-galactosidase gene
        - Renders it non-functional
- Recombinant plasmid is inserted into the antibiotic-sensitive bacteria by transformation
- Recombinant bacteria are grown on a medium called X-gal, which contains the chosen antibiotic, and a substrate for Beta-galactosidase
    - Only cells that contain the plasmid vector will grow on this medium
        - Since they have the gene for resistance to the chosen antibiotic
    - If foreign DNA was not successfully inserted into the Beta-galactosidase gene, the gene will code for functional enzyme that will hydrolyze a component of the X-gal medium to produce a blue colored compound
        - Produces blue colonies when a bacterium is allowed to divide
        - Blue color signals that these cells are non-recombinant
    - If foreign DNA was successfully inserted into the Beta-galactosidase gene, it will be non-functional and colonies of these cells will appear white on the X-gal medium

- Even with the successful production of recombinant cells, it is still not known if the desired gene of interest was inserted into the Beta-galactosidase gene, or whether some other DNA fragment was inserted (further testing is therefore required)
  - If the gene of interest codes for the production of an identifiable product, the bacterial isolate only needs to be grown in culture and tested
  - Otherwise, the gene itself must be identified in the bacterium via a procedure known as colony hybridization

Colony Hybridization

- DNA probes are short sequences of single-stranded DNA that are complimentary to the desired gene
  - DNA probes are synthesized in the lab
    - Some contain fluorescent dye
    - Some contain radioactive phosphorous
- DNA probes bind with the gene of interest and serve as markers for colonies that contain the gene of interest

Applications of Genetic Engineering

Therapeutic Applications

- Subunit vaccines
  - Nonpathogenic viruses carrying genes for a pathogen's antigens
- Gene therapy is used to replace defective or missing genes

Human Genome Project

- About 3 billion nucleotides make up the DNA of a typical adult human cell
- These nucleotides have been sequenced and all genes have been mapped
- May provide diagnostics and treatments by determining all possible proteins that can be produced

Sequencing Genomes

- Random Shotgun Sequencing
  - Small fragments of an organism's genome are sequenced
  - Then a computer is used to assemble the sequences in the proper order

- Computer software exists that can find the protein encoding regions of the sequenced DNA
  - Which can then be isolated and used in additional recombinant experiments
- Computer assisted DNA sequencing has led to a new field of study, bioinformatics (science of studying how genes function)
  - Proteomics is the science of determining all of the proteins a cell can express

Genetic Screening

- Recombinant DNA technology is also used in genetic testing for the presence of genetic diseases
- Southern blotting is a technique that can be used in the genetic screening process
  - DNA containing the gene of interest is removed from the cell and cut into pieces with restriction enzymes
  - Fragments are separated using gel electrophoresis
  - Fragments are then transferred to a filter by blotting
    - NaOH is used to separate the DNA fragments into single stranded molecules
    - Fragments on the filter are then exposed to a radioactive probe made from the defective form of the cloned gene of interest
    - Probe will hybridize with the defective gene if it is present in the sample of DNA taken from the cell
  - Radioactive probes that form hybrids with the defective gene are detected by exposing the filter to X-ray film
  - This technique can be used to test any person's DNA for the presence of a known defective gene
    - There are several hundred known defective genes that cause genetic diseases

Agricultural Applications

- Increase crop yields
- Make the plants resistant to herbicides, pests, drought, frost, viral infection, etc.

- Make the plants produce insecticides that kill insect predators

- Increase the shelf-life of fruits and vegetables after harvest

- Fix atmospheric N2, to reduce the amount of fertilizer that is applied to crops

Introduction to Genomics

Genomics is concerned with the structure, function, evolution, and mapping of genomes.

- Structural genomics is concerned with characterizing the structures of genomes

- Functional genomics attempts to describe gene, as well as protein, functions and interactions

- Proteomics is concerned with studying all the proteins encoded by the genome and their function

Cytogenetic Mapping

Cytogenetics is a branch of genetics that focuses on the study of inheritance in relation to the structure and function of chromosomes. A cytogenetic map is the visual appearance of a chromosome when it is stained and examined under a microscope. A cytogenetic map is used to determine the location of a particular gene relative to an observed banding pattern.

Linkage Mapping

Genetic linkage is the tendency of alleles located close together on a chromosome to be inherited together during meiosis.

- Linkage map – genetic map of species showing the position of their genes or genetic markers relative to each other

    o Relies on the frequency of recombination within species

- Molecular markers can be useful in mapping

    o Molecular markers are DNA segments uniquely recognized because of their association with a certain location within a genome

        ▪ Molecular marker characteristics vary among individuals

Physical Mapping

- Physical map – map of a chromosome or genome showing the physical locations of genes and other important DNA sequences

    o Requires cloning many pieces of chromosomal DNA

    o Cloned DNA fragments are characterized by:

      ▪ Size

      ▪ Gene components

      ▪ Relative locations along a chromosome

Human Genome Project (HGP)

- From HGP we learned

    o 3 billion base pairs make up one set of chromosomes

    o 25,000 genes

    o 1.5% genome encodes proteins

    o 96% similarity with chimpanzee genome

    o 0.17% of human genome varies between individuals

- Mutations in only 53 genes linked to disease when HGP began

    o Now more than 3,000 genes are linked to disease

- Information from HGP has led to improvements in diagnosis, treatment, and prevention of disease

CONCLUDING REMARKS

I hope this book has provided you tremendous value for your money and has helped you do better on your exams! If it has done both of these things, I have achieved my purpose in making this guide.

Furthermore, my goal is to create more books and guides that continue to deliver great value to readers like you for little monetary costs. Thank you again for purchasing this study guide and I wish you the best on your future endeavors!

- Dr. Holden Hemsworth

Your reviews greatly help reach more students. If you found this book helpful, please leave a review on Amazon, nothing helps more than a few kind words.

More Books By Holden Hemsworth

Do You Need Help with Other Classes?

Check out Other Books in the Ace! Series

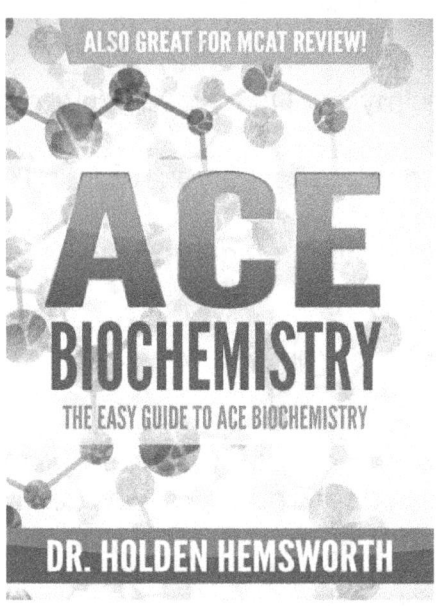

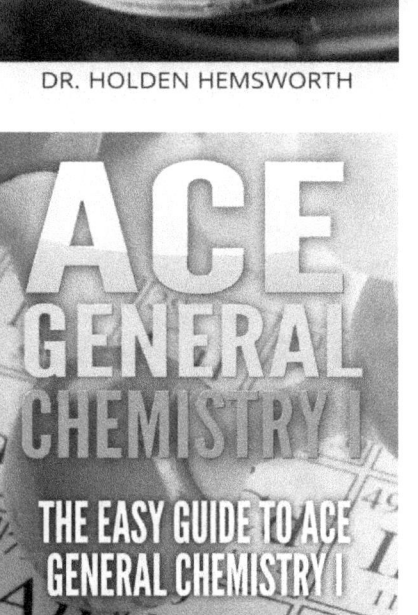

All Books are Listed on My Amazon Author Page

More Books Coming Soon!

www.ingramcontent.com/pod-product-compliance
Lightning Source LLC
Chambersburg PA
CBHW080813180526
45168CB00006B/2435